RANCHO LA BREA

A RECORD OF PLEISTOCENE LIFE IN CALIFORNIA

Assistant vertebrate paleontologist J. W. Lytle in the "bone room" in the basement of the Natural History Museum of Los Angeles County sorting material recovered from the 1913–1915 excavations.

RANCHO LA BREA

A RECORD OF PLEISTOCENE LIFE IN CALIFORNIA

BY
CHESTER STOCK

REVISED BY
JOHN M. HARRIS

SEVENTH EDITION

NO. 37
SCIENCE SERIES
NATURAL HISTORY MUSEUM
OF LOS ANGELES COUNTY

Natural History Museum of Los Angeles County
Los Angeles, California 90007
First Edition Published 1930. Seventh Edition 1992.
ISSN 0079-0943
Printed in the United States of America

CONTENTS

LIST OF ILLUSTRATIONS

FOREWORD TO THE SEVENTH EDITION

The continued popularity of Chester Stock's classic and authoritative work on Rancho La Brea has exhausted eight printings of the latest (sixth) edition. Major advances have occurred in the science of paleontology since the first edition in 1930, and subsequent work at Rancho La Brea has provided fresh insight into the diversity of the fossil biota and its geologic and ecological context. Nevertheless, much of Stock's original description and interpretation of remains from this unique site has withstood the test of time.

The seventh edition is based upon the text of the last edition compiled by Stock himself (1949). The scientific names for fossil and living animal species have been updated and additional information supplementing that provided by Stock has been integrated into the main portion of the text. Also added is a brief biography of Chester Stock, an updated bibliography, and a current list of the fossil plant and animal species recovered from the asphalt deposits.

Chester Stock drew freely on the results obtained by a large group of researchers who concerned themselves with Rancho La Brea and its fauna and flora. During the first half of the current century, John C. Merriam, in particular, furnished valuable contributions to many aspects of the Rancho La Brea occurrences including detailed studies of the Pleistocene mammals. The studies of Loye Miller likewise established a very substantial body of facts concerning the birds of the asphalt. Ida DeMay collaborated with Stock in the preparation of the second edition of this work published in 1942. Later editions reflected the increase in new material available for investigation and improvement of the original illustrations. Eugene J. Fischer was responsible for the construction of most of the mounted skeletons of extinct Rancho La Brea animals originally displayed at the Natural History Museum of Los Angeles County but these were subsequently improved and remounted by Leonard C. Bessom and Christopher A. Shaw prior to their present installation at the George C. Page Museum.

This revision owes much to knowledge and expertise freely shared by present and former members of staff, honorary associates, and volunteers of the Rancho La Brea section at the Page Museum. Particular thanks are due to William A. Akersten, Wesley Bliss, Kenneth E. Campbell, Jr., Shelley M. Cox, Sherri Gust, Lesley Marcus, George T. Jefferson, Richard Lamb, Cathy McNassor, Richard Reynolds, Mary Romig, Eric Scott, Christopher A. Shaw, and Antonia Tejada-Flores. The manuscript was kindly reviewed by Messrs. Shaw, Jefferson, and Marcus, and by Daniel M. Cohen and Craig C. Black. Assistance with the taxonomy of extant species was provided by Robert L. Bezy, Dan Cohen, C. Clifton Coney, Kimball L. Garrett, Sarah B. George, Robert J. Gustafson, Charles L. Hogue, James H. McLean, Jeff Seigel, Roy R. Snelling, Cindy Webber, Camm C. Swift, and Gary D. Wallace.

Original negatives from the Stock archives were reprinted by John S. Sullivan. Other illustrations were prepared by staff of the Exhibitions and Photographic Sections of the Natural History Museum of Los Angeles County.

This edition is issued in celebration of the one hundredth anniversary of the birth of Chester Stock.

John M. Harris
Earth Sciences Division
Natural History Museum
of Los Angeles County

Figure 1. Chester Stock examining dire wolf skulls and teeth (1939).

CHESTER STOCK

(January 28, 1892–December 7, 1950)

During his lifetime, Chester Stock was regarded as one of the most important and respected mammalian paleontologists in the United States. His writings on the subject of Pleistocene life in North America are among the more significant contributions in this topic during the first half of the 20th century and remain widely used sources of reference.

Chester Stock was born to German immigrant parents in San Francisco, California. His early life was spent in one of the poorer parts of the city, but Stock excelled in school and, as a young child, became fascinated with science. After his family's finances were devastated by the 1906 San Francisco earthquake, Chester left high school to work at the Old Union Iron Works. He was not a particularly robust young man and the hard labor adversely affected his health. In 1910, after he had recovered, Stock entered the University of California at Berkeley, where his future career took shape. He became a favorite student of the eminent paleontologist John C. Merriam and was persuaded to pursue this field rather than to study medicine. Merriam steered Stock's interest toward the study of Pleistocene faunas and assigned him projects on some of the most exciting finds of the day. One such assignment, ground sloths from Rancho La Brea, became the topic of Stock's master's thesis. He earned his doctorate at Berkeley in 1917 for a dissertation on Hawver Cave, a paleoanthropological site in El Dorado County, California, with an associated Pleistocene fauna.

Upon graduating, Stock entered the faculty at Berkeley as an instructor. Merriam left Berkeley in 1921 to assume the presidency of the Carnegie Institute, but the continued association of the two men contributed to a significant outburst of research activity on the West Coast. The Carnegie Institute funded many of Stock's projects in the ensuing years, including exploration of the Oligocene John Day Formation in Oregon, excavation of Pleistocene asphalt seeps at McKittrick, California, and continuing research on Rancho La Brea. During his years at Berkeley, Stock wrote many articles on a diversity of subjects and, in 1925, the Carnegie Institute published his major work on the ground sloths— *Cenozoic Gravigrade Edentates of Western North America, with Special Reference to Megalonychinae and Mylodontidae of Rancho La Brea.*

In 1926 Chester Stock and John Buwalda became the founding members of the Geology Department at the California Institute of Technology (Caltech) in Pasadena, California. Through student and professional excavations, aided by some purchases, Stock built an impressive collection of vertebrate fossils at Caltech, which swiftly became one of the major centers for paleontological research in the country. Stock's research activity continued unabated, but he also became a popular professor and a much sought-after public lecturer.

His relocation from Berkeley to Los Angeles provided Stock with greater access to the huge collection of fossils excavated from Rancho La Brea by the Los Angeles County Museum between 1913 and 1915. His overview of assemblages from Rancho La Brea—the first edition of the present volume (1930)—was the first scientific work published by the museum. In 1932 Stock and Merriam completed their joint monograph *The Felidae of Rancho La Brea*, which is still considered a masterpiece of paleontological research.

Stock's interest in the collections of the Los Angeles County Museum led to his appointment as Senior Curator in charge of the Earth Sciences section, and he served as Chief Curator of the Division of Science during the last two years of his life. In this capacity he was instrumental in launching the program of development in Hancock Park that later led to recognition of Rancho La Brea as a major scientific monument, for which it was accorded National Natural Landmark status.

Stock initiated many other field projects, some of the more notable being excavations in the Eocene Sespe Formation, on the California Channel Islands, and at San Josecito Cave in northern Mexico. He was very interested in the early occurrence of humans in the New World, and was frequently consulted about the identifications of extinct animals associated with the remains of paleo-indians. To this end, he worked with archaeologists on faunas from Gypsum Cave in Nevada and Sandia Cave and Clovis sites in New Mexico.

Despite the diversity of his other research interests, the animals from Rancho La Brea retained an enduring fascination for Stock until his untimely death in 1950, and provided the topics for some of the most important of his 170 publications. Much of our current understanding of the nature and context of the faunas of the Los Angeles asphalt seeps was obtained through Stock's pioneering investigations, and his reseach will continue as a lasting contribution to the study of terrestrial fossil biotas.

Stock's collection of scientific reprints forms the nucleus of the Chester Stock Memorial Library of the George C. Page Museum, and much of his correspondence is on file in the Stock Library archives. In 1951 George Gaylord Simpson published a detailed biography of Stock (*National Academy Biographical Memoirs* 27: 335–362). Most of the vast collection of vertebrate fossils acquired by Caltech under Stock's direction was purchased by the Natural History Museum of Los Angeles County in 1957, forming the nucleus of that museum's vertebrate paleontology holdings.

Cathy McNassor
Rancho La Brea Section
George C. Page Museum

INTRODUCTION

The unique collection of fossils obtained from the asphalt deposits of Rancho La Brea has no parallel among the numerous records of the past life of the earth brought to light by paleontologists and geologists. Rancho La Brea is the best known of three fossiliferous asphaltic deposits of similar age from California, the others being located at Carpinteria, Santa Barbara County, and at McKittrick, in Kern County. Dating from a period not very remote in earth history, yet possessing considerable antiquity as measured in terms of years, the collection from Rancho La Brea furnishes a basis for reconstructing a remarkably informed picture of life as it existed in the Los Angeles region of Southern California between 4,000 and 38,000 years ago.

Among the outstanding features of the Rancho La Brea collection are the great wealth of material, the unusual variety of the species, and the fine state of preservation of the remains. Together, these provide an incredibly detailed source of information about the animals and plants preserved as fossils at this locality. The mounted skeletons of many of the characteristic mammals and birds displayed in the George C. Page Museum of La Brea Discoveries, a satellite facility of the Natural History Museum of Los Angeles County, are only a small part of the collection obtained from the asphaltic deposits on which the Page Museum itself is built. In many instances, individual skulls and parts of skeletons are duplicated many times by specimens not on exhibition. Represented by more than 2,000,000 bones, 100,000 insect fossils, 40,000 mollusks, and a large but undetermined quantity of plant and microscopic animal remains, more than 600 different kinds of animals and plants are now known from Rancho La Brea. To this list other forms will inevitably be added as the study of the entire assemblage progresses.

In 1951 the assemblages of fossil mammals from Rancho La Brea were selected to characterize the Rancholabrean Land Mammal Age (Savage, 1951), thus formally recognizing the significance of Rancho La Brea to our understanding of life on this continent during the latest portion of the Pleistocene Epoch. In 1963 Hancock Park, which contains the Rancho La Brea site, was declared a National Natural Landmark in recognition of its importance to science. In 1974, *Smilodon californicus*, whose type locality is Rancho La Brea, was declared the official state vertebrate fossil of California, although *Smilodon californicus* is now more correctly known as *S. fatalis*.

It is not surprising, therefore, that the occurrence and collection have aroused considerable interest on the part of both the scientific specialist and the layman. Much intensive research during the past forty years or more has resulted in the accumulation of a fund of information relating to these deposits and their exhumed organic remains. The present review of our knowledge of these deposits and of their record of life is intended to further this interest and to serve the needs of the visitor to the Page Museum and the site of Rancho La Brea.

HISTORY OF DISCOVERY AND DEVELOPMENT

The Rancho La Brea "Tar Pits," located on the north edge of Wilshire Boulevard a few miles west of the city center, have long served as a major tourist attraction, drawing scientists and lay visitors from all over the world, as well as having provided a seemingly endless source of jokes.

The name stems from the time of the 1828 Mexican land grant of Rancho La Brea (literally "Tar Ranch"). This name originally applied to the entire 4,500-acre land grant, but in modern usage refers to the plot of ground in which the fossil-bearing beds are located. Strictly speaking, the use of "tar" or "brea" in this context is inaccurate. The naturally occurring bituminous substance derived from petroleum is asphalt, whereas tar is a commercial product obtained from the fractionation of petroleum. At Rancho La Brea the asphalt wells up from an oil field that lies deep below the ground surface (Figure 2). The material proved useful to the early settlers of Los Angeles, and the original land grant stipulated that residents of the town should have the unimpeded right to carry away any asphalt needed to waterproof the roofs of their houses. Later the site was mined commercially, most intensively in the late 1860s and early 1870s, for use in road construction and the building industry.

Asphalt seeps or "springs of pitch" in the Los Angeles region were apparently first recorded by Gaspar de Portolá in his diary of the Portolá California Expedition of 1769–1770 as follows: "The 3rd [August 3, 1769], we proceeded for three hours on a good road; to the right of it were extensive swamps of bitumen which is called *chapapote*. We debated whether this substance, which flows melted from underneath the earth, could occasion so many earthquakes."

A second early report of the occurrence was made by José Longinos Martínez in the journal of his expedition to California in 1792 (see Simpson, 1961). He stated that: "Near the Pueblo de Los Angeles [there are] more than twenty springs of liquid petroleum, pitch, etc. To the west of the said town, in the middle of a great plain of more than fifteen leagues in circumference, there is a large lake of pitch, with many pools in which bubbles or blisters are constantly forming and exploding. . .In hot weather animals have been seen to sink in it, unable to free themselves because their feet were stuck, and the lake swallowed them. After many years their bones come up through the holes, as if petrified. I brought away several specimens of them.

"For a great distance around these volcanoes there is no water, and when the heat of the sun forces birds to seek it they alight upon the lake, mistaking it for water. All the birds that do so are caught by the feet and wings until they die of hunger and thirst. Rabbits, squirrels, and other animals are deceived in the same way, and for this reason the gentiles keep a careful watch at such places in order to hunt without effort. Off the road near San Buenaventura there are several other springs of bituminous petroleum, and near them deposits of the same pitch, hardened."

The French explorer Eugène Duflot De Mofras made the following statement in the account (1844) of his explorations in Oregon and California: "Two leagues to the southeast

Figure 2. View looking northwest showing portions of Rancho La Brea and the Salt Lake Oil Field with the Santa Monica Mountains in the background. Exploratory excavations for fossils shown with Pit 4, flooded, in middle foreground. Photograph taken February 1914; Page Museum Archive.

of Los Angeles there are four great sources of asphaltum, situated on a level with the earth in a vast prairie. These springs open in the middle of little pools of cold water, while the bitumen possesses a higher temperature. This water has a mineral taste, which, however, does not prevent animals from drinking it. At sunrise the orifices of these springs are covered by enormous bubbles of asphaltum, often being more than a yard high, and looking like soap bubbles." On the map accompanying his report De Mofras indicates the source of the bitumen in the plains west of Los Angeles.

The first cartographic record of the position of bituminous springs, at Rancho La Brea, was that made by E. O. C. Ord in 1849. On a topographic sketch map of the Los Angeles plains and vicinity, issued with Lieutenant Ord's report, the location of pitch springs is shown at a point several kilometers west of Los Angeles and south of the gap in the mountains now known as Cahuenga Pass. In 1853 the Los Angeles region was again examined as a part of the program of exploration for a railroad route from the Mississippi Valley to the Pacific Coast. In the report of this expedition the geologist William P. Blake described (1856) the occurrence of bituminous deposits, possibly those of Rancho La Brea

It was not until 1875, however, that a published statement referred to the occurrence of skeletal remains of extinct animals in the asphalt deposits of Rancho La Brea. In that year, William Denton of the Boston Society of Natural History gave an account of his visit to the brea ranch of Major Henry Hancock and described the asphalt accumulations which were then being excavated for their asphalt content. He stated furthermore that Major Hancock presented him with a tooth which was later determined to be a canine tooth of a saber-toothed cat. Denton at the time of his visit also secured bones and teeth of a fossil horse and of other mammalian and bird remains.

The account given by Denton apparently escaped further notice, and no other interest in the occurrence of fossil materials at this locality was expressed for another thirty years. The next person to recognize the importance of the fossil bones and teeth in the deposits was the geologist W. W. Orcutt of Los Angeles. In 1901 Orcutt visited the locality to determine the feasibility of oil production. His interest piqued by the several fossil bones he collected on that visit, he sought the permission of Madame Ida Hancock Ross to continue his visits over the next several years. In 1906, when his collection had grown to include a portion of a sabertooth skull, jaws of a large wolf, and dermal bones of a large ground sloth, he placed this material at the disposal of Dr. John C. Merriam of the University of California. Dr. Merriam appreciated the significance of this discovery and after visiting the locality was convinced that further excavation would yield larger collections.

Permission to excavate was granted the University of California by Madame Hancock Ross and subsequent explorations were carried out at intervals from 1906 to 1913. The Southern California Academy of Sciences, Occidental College, and the Los Angeles High School also obtained collections during that period.

In 1913 G. Allan Hancock granted Los Angeles County the exclusive privilege to excavate at Rancho La Brea for a period of two years. The excavations were conducted by the Los Angeles County Museum and the materials that were obtained are accessioned in the Natural History Museum as the Hancock Collection, a memorial to Major Henry Hancock and Madame Ida Hancock Ross. During this two-year interval, hundreds of thousands of bones were excavated from the asphaltic sediments

In May 1915, Mr. Hancock offered to give the tract of land on which the famous fossil beds occur, approximately 23 acres, to Los Angeles County with a request that the scientific features of the site be adequately exhibited and preserved. A public park, known as Hancock Park, was subsequently established. The park faces Wilshire Boulevard on the south and is bordered on the west by Ogden Drive, on the north by Sixth Street, and on the east by Curson Avenue. The remains of several of the original excavations made in search for fossil material can still be seen. In the mid-1940s, plans for relandscaping the park were prepared by an architect, and work was begun several years later. Included in this project was the erection of a building over a fossil occurrence in the western part of the park. This observation station permits the visitor to descend by a circular staircase to a promenade from which may be seen the bones and skulls of the ancient animals that were trapped and entombed during Pleistocene time, and subsequently brought to view again by paleontologists. Placed elsewhere within the park were life-sized restorations of several of the characteristic animals recovered from the Rancho La Brea deposits, some fabricated under the direction of Chester Stock.

In 1969, excavation resumed at Pit 91, close to what is now the north entrance to the Los Angeles County Museum of Art, in order to further investigate the mode of accumulation of this and other asphalt deposits and specifically to recover the remains of microscopic animals and plants that had been preserved (Shaw, 1982). Year-round excavation took place until 1980 and today continues during the summer months when visitors may view the excavation in progress.

Chester Stock had strongly believed that an on-site museum would best serve the educational function of the fossil collection. Construction on the long-anticipated museum was finally begun in the eastern part of Hancock Park in 1975, made possible by funds donated by George C. Page. In 1977 the George C. Page Museum of La Brea Discoveries

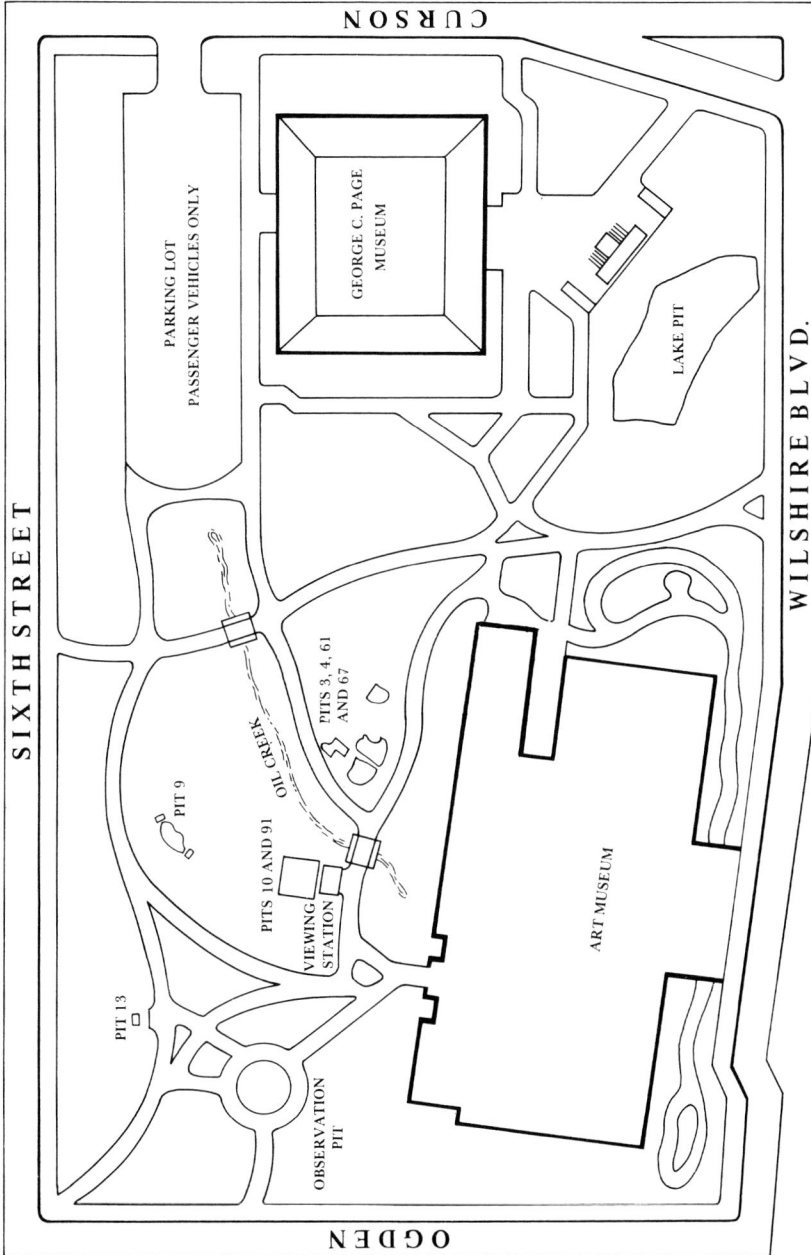

Figure 3. Map of Hancock Park

Figure 4. Ground squirrel (*Citellus*) mired in Recent asphalt seep at Rancho La Brea.

was opened (Figure 3). The museum displays examples of the variety of fossil materials recovered from the asphalt deposits and houses the extensive Hancock collection of fossils. Fossil materials collected by the Southern California Academy of Sciences at the turn of the century and those salvaged from construction activities in and around the periphery of the park are also housed in the Page Museum. An additional viewing platform erected at the shore of the lake at the southeast corner of the park affords spectacular views of escaping methane gas. The lake, however, is not the "great lake of pitch" referred to by José Longinos Martínez but constitutes the flooded excavation sites of former commercial brea mines.

Seepage of the asphalt, and occasional entrapment therein of small animals, is still in progress at the present time (Figure 4). While the activity which has brought about the surface outpours may have diminished considerably since the time of formation of the bone-bearing asphaltic deposits, the accumulations forming today give an impressive demonstration of the conditions and processes that prevailed during an earlier time. Thus, within the corporate limits of metropolitan Los Angeles there still remains at Rancho La Brea an indubitable link with the reality of the geologic past.

POSITION IN GEOLOGIC TIME

The first scientific investigations of Rancho La Brea fossils took place long before the development of modern radiometric dating techniques. Hence, their geologic age was evaluated from the stratigraphic relationships of the fossil-bearing deposits and by comparing the Rancho La Brea assemblages with others that had been discovered elsewhere in North America. Stock's interpretation that the Rancho La Brea deposits were formed during the latest portion of the Pleistocene epoch was subsequently confirmed and refined by radiometric dates derived from fossil bone and wood.

The asphaltic beds are essentially part of a sedimentary series consisting of sands, clays, gravel, and angular rubble whose present thickness is between 12 and 58 m (40 to 190 ft) in the area of the old Salt Lake Oil Field, lying immediately to the north of the fossil beds (see page 11). Beneath these sedimentary deposits are older formations of marine shales and sandstones with interbedded oil sands from which the petroleum has come. The attitude and relationships of these older beds clearly indicate that they were folded and to some extent eroded prior to the accumulation of the more or less flat-lying strata containing the Rancho La Brea asphalt deposits. Obviously, the earliest age which may be assigned to the latter strata is determined by the age of the latest beds affected by the folding and subsequent erosion.

Geological and paleontological investigations demonstrated that not only are later Tertiary marine strata folded in the Los Angeles Basin area, but early Pleistocene deposits are likewise deformed. In other words, the earth movements which brought about the folding of the older strata and the extensive erosion which followed this episode occurred within Pleistocene time and mark an important break between the strata that accumulated before and after these events. The older beds, as Eaton (1928, p. 134) pointed out, are everywhere deeply eroded and seldom show a gently sloping or horizontal attitude. In contrast, the later deposits with the asphalt lenses at Rancho La Brea retain their essentially horizontal position of accumulation. The Rancho La Brea deposits thus have to be younger than the deformed early Pleistocene strata.

Grant and Sheppard (1939, p. 308) suggested that the older alluvium containing the asphalt beds represents marginal deposits of the Hollywood alluvial fan. This fan is one of several similar fans forming the piedmont slope along the southern border of the Santa Monica Mountains. In the western part of this region these fans are now being dissected by streams, a feature of the present cycle of erosion. The absence of any considerable thickness of sediments overlying the horizon of the asphalt beds in the vicinity of Rancho La Brea would seem to indicate that the conditions of accumulation have remained similar to those found today at this locality. In their geological studies of the Palos Verdes Hills, San Pedro, located 32 km (20 mi) south of Rancho La Brea, Woodring, Bramlette, and Kew (1946) pointed out that the alluvium in which the fossiliferous brea accumulations occur is essentially the same age as the older alluvium which is now arched over the Dominguez Hills some 19 km (12 mi) to the south. This older alluvium was regarded by them as the equivalent of the non-marine cover of the lowest marine terrace on the face of the Palos Verdes Hills. Considerations of this kind confirm the view that the fossiliferous asphalt was laid down in later Pleistocene time.

The Rancho La Brea fossils were thus preserved in gravels and sands accumulating on the outwash plain between the Santa Monica Mountains and the Pacific Ocean. The stratigraphy of the deposits has been summarized by Woodard and Marcus (1973, 1976) and Shaw and Quinn (1986) but no formal rock unit names have been proposed (Figure 5). The main bone accumulations are restricted to the upper 9 m (30 ft) of sediments, and are underlain by a bed of sand containing marine fossils and representing the last recorded marine incursion in the area (Marcus and Berger, 1984). Similar marine fossils typify the Sangamonian Interglacial—about 100,000 years before present (Valentine and Lipps, 1970; Woodard and Marcus, 1973). Distinctive rock types found as clasts in the alluvial deposits suggest that, in the vicinity of Rancho La Brea, the outwash plain had a northwest source area during the late Pleistocene (Benedict Canyon, Coldwater Canyon, etc.) and was tectonically uplifted in the early Holocene (Quinn, 1991). The asphalt itself has its origin in petroleum from the Salt Lake Oil Field and the widespread distribution of asphalt seeps

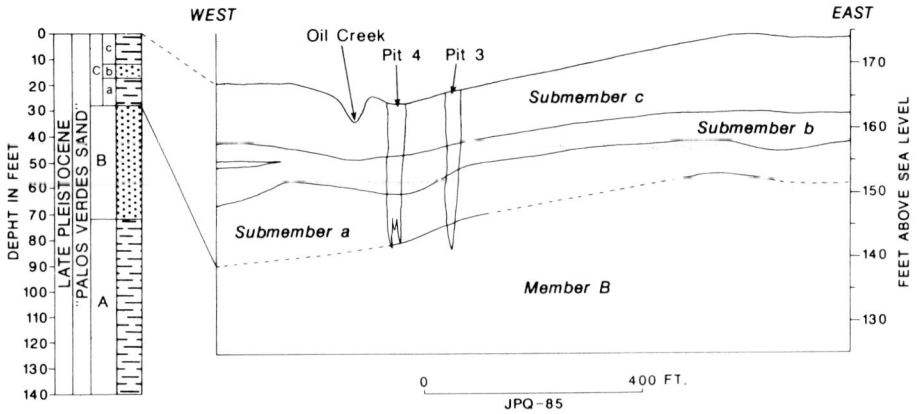

Figure 5. Stratigraphic column and cross section of the fossiliferous strata at Rancho La Brea. The cross section is aligned in a general east–west direction across Hancock Park. After Shaw and Quinn (1986).

suggests extensive subsurface fracturing during the formation of the flexure responsible for the oil field (Arnold, 1907a). Many of the fossiliferous sites within Hancock Park are clustered along an axis that traverses the park in a northwest-southeast direction, suggesting that the asphalt springs may originate from a subsurface (Sixth Street) fault.

From a paleontological perspective, many of the larger animals fossilized at Rancho La Brea are similar to or identical with species described elsewhere from Pleistocene horizons. However, the plants and many of the smaller fossil animals from this locality belong to species that are still living today. This combination of living and extinct species indicates that the Rancho La Brea fossils represent an interval late in the Pleistocene Epoch (Figure 6). The presence of many extinct animals suggests that widespread extinction occurred in the animal world after the entombment episodes at Rancho La Brea. If this episode of extinction occurred in the later part of the Pleistocene, it must have been of relatively short duration.

The fauna obtained from the excavations of the Natural History Museum of Los Angeles County appears to be essentially a homogeneous one. To be sure, certain kinds of mammals are better represented in some pits than in others, but the possibility of a catastrophe overtaking an entire troop of mammoths or a family of ground sloths in one seep can not be wholly disregarded. Different elements present in the fauna suggest different ecological or environmental factors. Thus, the mastodont, the ground sloth *Megalonyx*, peccaries, deer, and timber wolves seem to indicate forest conditions and therefore reflect an environment different from that normally occupied by the ground sloth *Glossotherium*, camel, bison, horses, and many other mammals. It has yet to be established whether this admixture reflects the prevailing disposition of available habitats or environmental fluctuation affecting the region through time.

Radiometric dates have been obtained from wood and bone from the Rancho La Brea deposits using a variety of different methods (see Table 1). Radiocarbon dates utilizing collagen appear to be more accurate and consistent than those based on other bone components (Ho et al., 1969; Marcus and Berger, 1984). McMenamin et al. (1982) found that amino acid racemization rates were quite variable but were, overall, significantly slower in Rancho La Brea specimens than in those from more typical sedimentary deposits. Bischoff and Rosenbauer (1981) found agreement between uranium series dates and collagen ^{14}C ages of *Smilodon* femora from the same level in the same locality. The oldest dates on bone are about 32,000 years BP (before the present) and the youngest are less than 6,000

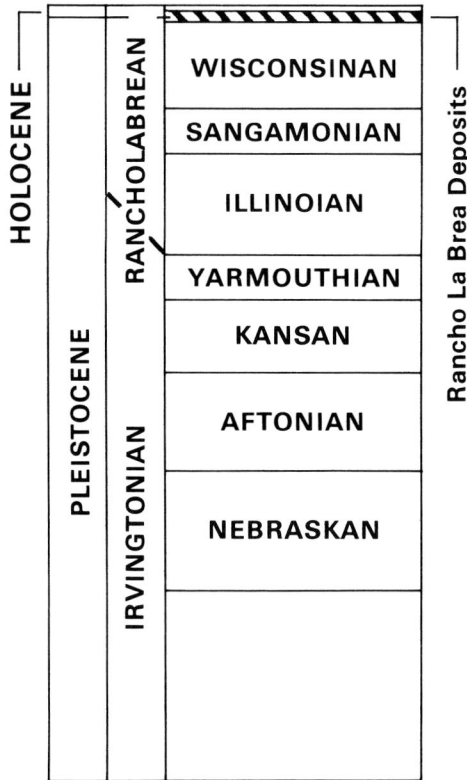

Figure 6. Diagram showing the chronologic position of the Rancho La Brea deposits. The left column shows geological epochs, the center column North American Land Mammal Stages, and the right column glacial and interglacial stages. The fossiliferous asphaltic strata accumulated at the end of the Wisconsinan glacial and are of latest Pleistocene and earliest Holocene age. The fossil assemblages are typical of the Rancholabrean land mammal stage but include a number of Recent (Holocene) species.

Table 1. Estimated age of fossil deposits from different Rancho La Brea localities. The figures constitute the average dates derived from samples of bone collagen, except those for Pit 9 which were derived from fossil wood. Data from Marcus and Berger (1984).

Location	Average fossil age (years)
Pit 3 (upper levels)	14,000
Pit 3 (lower levels)	20,000
Pit 4	24,500
Pit 9 (upper levels)	13,500
Pit 9 (lower levels)	38,000
Pit 13	15,000
Pit 16	23,000
Pit 60	26,000
Pit 61/67	12,000
Pit 77	31,000
Pit 81	11,000
Pit 91	29,500
Pit 2051	21,000

years BP. Thus, the Rancho La Brea biota became fossilized during the glacial maxima and interstadials of the last (Wisconsinan) glaciation of the Pleistocene Ice Age. No extinct species have been recovered from sites interpreted as less than about 11,000 years old, and such younger deposits have yielded mainly the remains of small mammals and birds together with some human artifacts. The presence of pipes, chimneys, and pockets filled with asphalt containing the bones of Holocene mammals and birds is not unexpected in view of the constant movement of gas and oil to the surface from the oil sands below and, indeed, the same process continues today.

CLIMATIC CONDITIONS

The occurrence of many large extinct mammal and bird species and the great diversification of the entire fauna may at first glance be taken as a satisfactory basis from which to infer a radically different climate than that of the present. However, many of the larger species, in contrast to those of the rodents and perching birds, are known to have ranged widely over the North American continent and especially through the southwest. The needs of the large herbivorous mammals would have been provided with only a slight increase in rainfall typical of that today, and the presence of aquatic plants, water insects and bugs, and other aquatic invertebrates, fish, and amphibians confirms the existence of standing water or slowly flowing streams for at least part of each year. The number and diversity of the herbivores would in turn have supported the quantity and diversity of the large carnivores preserved in the deposits. It seems significant that creatures like the tapir, which today live in moist savannas of the tropics, are very rarely represented in the fossil record at Rancho La Brea. On the other hand, the absence of forms that are associated elsewhere with cold climates, musk-oxen for example, suggests that conditions were not rigorous.

It should be remembered that the smaller mammals (rodents, rabbits, and shrews) often have limited geographic ranges and thus strongly reflect local environmental conditions. Many of the smaller vertebrates and invertebrates from Rancho La Brea belong to species which persist today in the same region. This suggests, therefore, no great difference in climate of the Los Angeles Basin between the time that they were entombed and the present.

The fossil plants recovered from Rancho La Brea are of even greater potential significance in this connection because plants can be very sensitive indicators of climate. Unfortunately, the plant record from the asphalt deposits of Rancho La Brea is not yet fully studied. However, four distinct associations appear to be represented—coastal sage scrub, riparian woodland, chaparral, and deep canyon floras. Johnson (1977a, b) suggested that "full glacial winters in coastal California may have resembled modern winters but summers may have been cooler and perhaps more moist than now (but still relatively dry)." Lamb (1989), on the basis of the fossil mollusks, interpreted annual precipitation in the Los Angeles Basin during the Late Pleistocene to be twice that of today. Carbon[13] analyses of bone collagen provided $\delta^{13}C$ values in the range of C_3 plants or plant consumers, suggesting summer temperature minima were cooler than those of the present day (Marcus and Berger, 1984).

The interval of time represented by the Rancho La Brea deposits witnessed considerable temperature fluctuation elsewhere on the continent. Preliminary studies have detected temporally related changes in overall body size in some of the larger fossil mammal species (Menard, 1947; Nigra and Lance, 1947; Shaw and Tejada-Flores, 1985) that may reflect climatic changes in the Los Angeles Basin during the late Pleistocene. This might, in turn, help explain apparent inconsistencies between indications of environment provided by different elements of the biota. However, detailed investigation of the composition and character of the fauna and flora in different parts of the sequence has yet to be undertaken.

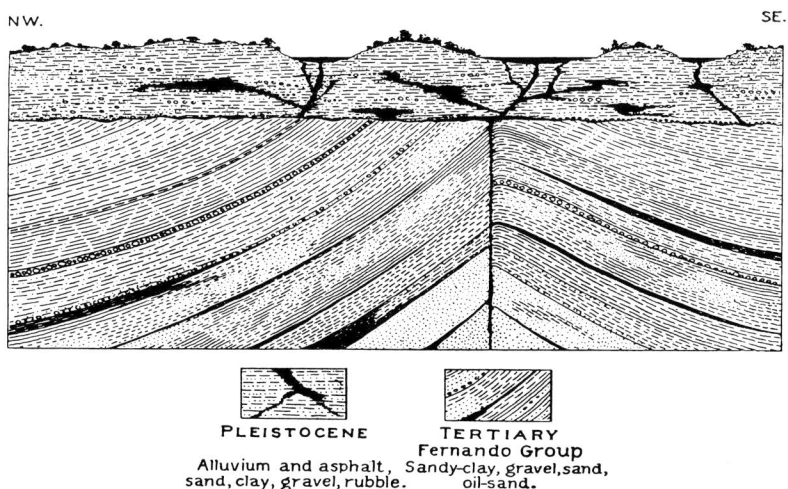

PLEISTOCENE   TERTIARY
Fernando Group
Alluvium and asphalt, Sandy-clay, gravel,sand,
sand, clay, gravel, rubble.   oil-sand.

Figure 7. Generalized cross section showing geologic structure and relationships of formations at Rancho La Brea during period of miring of Pleistocene animals and plants. Character and structure of the sediments containing the oil sands taken from section in Salt Lake Oil Field; modified after Arnold (1907a).

PHYSICAL FEATURES AND ORIGIN OF THE ASPHALT DEPOSITS

It seems evident that the processes responsible for the accumulation of the fossiliferous asphalt deposits persist today at Rancho La Brea, although perhaps with lesser intensity. The source of the petroleum is the oil sands that are interstratified with the older shales and sandstones underlying the Pleistocene strata at Rancho La Brea. As determined by the geologic structure (see Figure 7) in the Salt Lake Oil Field, these older marine strata are deformed and folded. Immediately to the north of Rancho La Brea an upward flexure of the older rocks, whose crest has been broken, extends apparently in the northeast–southwest direction, and without much question facilitates the upward movement of gas and oil in the general vicinity of the asphalt beds. Subsidiary underground structures, as for example, local fractures of a minor fold in the older sediments beneath Hancock Park, may account for the apparent localization of the productive brea accumulation at the site of Rancho La Brea.

Penetration of the sedimentary strata by the petroleum, resulting in asphalt seeps and asphaltic material at ground surface, occurred concomitantly with accumulation of the Pleistocene alluvial deposits. At present, oil and gas reach the surface through small fissures, pipes, or chimneys, the oil forming small and generally shallow pools about the vents. Both here and in the artificial lake in front of the Page Museum, bubbles of gas rise constantly to the surface and the outpours of asphalt spread over and through the soil of the adjacent ground. An interesting aspect is the occasional downward movement of the asphalt that can be discerned at the vents. A temporary release of gas pressure below permits the asphalt to recede again into the pipe or chimney from which it has exuded. The downward flow may carry remains of organisms or hardened lumps of asphalt and other detrital materials from the surface into the pipes. This or a similar action may be responsible for the movement of submerged animal remains in the larger Pleistocene outpours. When the asphalt reaches the surface its more volatile constituents escape, leaving a denser residue which becomes crusted. This hardening of the surface may occur rapidly along the edges of the seep and may extend gradually inward towards the middle. The surface may remain exposed for some time or may be gradually covered by soil or dust. In the warmer seasons of the year or during the heat of the day the asphaltic crust becomes quite soft.

During the Pleistocene the exudation of the asphalt could have been much more extensive than at the present time. The pools of oil occupied the natural depressions of an irregular land surface and were on occasion several square meters in area. The depth and borders of these seeps were probably variable. Excavations conducted at Rancho La Brea have shown that the fossiliferous asphalt was frequently of irregular outline and varied from relatively shallow accumulations to thick deposits having a maximum depth of 9 to 11 m (30 to 36 ft). On the average, however, productive pits extended from a short distance below the surface to a depth of approximately 4 to 7 m (13 to 23 ft).

That asphaltic eruptions contributed to the building up of the land surface is suggested by interesting evidence secured in Pit 3 of the Natural History Museum of Los Angeles County excavations. Here, at a depth of just over 1 m (4 ft), an upright trunk of a juniper tree was encountered. The top of the tree had either burned or rotted away. At a depth of 2.2 to 2.6 m (7.5 to 8.5 ft) a large limb projected from the trunk, and at a level of 3.6 m (12 ft) below the present surface the tree was found to be rooted in a stiff clay. Its trunk was surrounded by clay, sand, and asphalt. Also packed around the limb and portions of the trunk were dense masses of bones and skulls representing many of the typical mammals and birds of the Rancho La Brea Pleistocene assemblage. There can be little question that the tree occurred *in situ*, perhaps originally growing along the border of a depression in which detrital materials and asphalt were accumulating. Continued exudation of oil may have caused the asphaltic mass to encroach upon the tree, with consequent juxtaposition of the entombed animal remains.

Not all of the seeps need have formed by the filling of natural depressions at the Rancho La Brea locality during the Pleistocene. Expulsions of large quantities of gas and oil may have produced craterlike vents several meters in diameter. Presumably such vents were later partially or totally filled by an inflow of viscous material. Moreover, the rim of a vent of this type might be elevated a short distance above the general level of the adjacent surface, forming a buttress against which clays, sands, and rubble were later deposited. Figure 8 illustrates one such "tar volcano" at the Carpinteria asphalt mine that was described by Arnold (1907b).

Additional information pointing toward this method of formation seems to be found in some of the productive fossil pits excavated by the museum. Contour records of some of these excavations indicate a more or less conical mass of asphalt in which fossil remains are preserved. However, the apparent "funnel shape" of many of the fossiliferous deposits may have owed more to the way in which they were excavated than to the original dimensions of the fossiliferous asphalt bodies—the quarry walls being sloped gently outward to minimize their subsequent collapse during the recovery of the fossils. The general geometry of these deposits does, nevertheless, reflect the concentration of fossils in asphalt pipes and adjacent surface flows (Shaw and Quinn, 1986). An exception to this "norm" was encountered during construction of the Page Museum when a thin, tabular fossiliferous accumulation was discovered in the area now occupied by that museum's atrium. This latter occurrence was also remarkable in that it contained a number of articulated skeletons (Jefferson and Cox, 1986), contrasting with the unassociated material that forms the bulk of the collections.

Interpretation of the age and geometry of the bone accumulations is rendered more difficult by the different ways in which these accumulations were formed. An apparently single bone accumulation may represent more than one temporal interval because of the

Figure 8. Example of a "tar volcano" in the Carpinteria Asphalt Mine, near Carpinteria, California. Photograph by Ralph Arnold; courtesy of U.S. Geological Survey.

episodic nature of the asphalt seepage. A new accumulation may form at a topographically lower elevation than adjacent older ones because of the cut and fill nature of fluviatile deposition, a situation which is further complicated by the fact that hardened and oxidized asphalt is greatly resistant to erosion. Conflicting radiometric dates for bones from the same locality may result from variability in the composition of fossil bone or natural but otherwise undetectable stratigraphic intermixing. It is clear that no single simple explanation will account for the variety of fossil occurrences or dates from the different localities; only the excavation at Pit 91 has been sufficiently well documented to permit, at some future date, an accurate reconstruction of accumulation at one part of the Rancho La Brea site (Marcus and Berger, 1984).

The visitor to the fossil-bearing deposits at Hancock Park must not construe the open pits, now on view there, as surface features that represent the active traps where animals were caught during the Ice Age. Quite to the contrary, these excavations serve only as a record of the explorations for fossil remains that were conducted, for the most part, nearly eight decades ago.

MODE OF ACCUMULATION OF THE FOSSIL MATERIAL

The small seeps of asphalt that now form at Rancho La Brea are known to catch and hold in their midst the unfortunate creatures who by chance come in contact with the sticky substance. Thus, the seeps present a unique and most efficient type of trap, operating almost unceasingly and capable of catching many of the birds, mammals, and insects now inhabiting the region. Similarly, the extensive Pleistocene outpours, situated in a region richly stocked with vertebrate life, also entrapped animals, but the larger size and conceivably greater depth of these seeps permitted even greater tragedies to occur. It is not difficult to visualize some of these Pleistocene catastrophes. A single animal, large or small, becoming mired in the asphalt would naturally lure others to the trap. The carnivorous birds and mammals seeking to reach this bait would frequently fall victims to the tenacious grip of the viscous material and thus in turn would serve to attract still other creatures to the seep. It is conceivable that a single seep might gather into its mass in a relatively short time a great many victims whose remains now form the remarkable accumulation of bones, skulls, and teeth found at Rancho La Brea. Nevertheless, because of the 30,000 or more years represented by the fossil accumulations, only one such major entrapment episode (involving, for example, a large herbivore, four dire wolves, a sabertooth, and a coyote) need have occurred every decade to account for the immense number of fossil mammals represented in the collections.

That the carnivores were inevitably attracted to prey fastened in the asphalt and were therefore particularly susceptible to entrapment and subsequent entombment is clearly attested by their preponderant representation. The victims of this process include young, aged, and maimed individuals. Doubtless the age of an individual animal, its keenness in sensing danger, and its ability to secure food in or away from the traps are but a few of the factors contributing to the nature of the specimens trapped in this type of accumulation.

The methods by which fossils accumulated within the asphaltic deposits have still to be thoroughly and satisfactorily explained (Akersten, 1991; Quinn, 1991). Domestic animals including cows are known to have been caught and immobilized by asphaltic outflows only a few centimeters in depth, and hence deep seeps of asphalt were not essential to the entrapment of the Rancho La Brea biota. The relative scarcity of articulated or associated skeletons also argues against wholesale entombment of complete animals, at least on a regular basis. Birds (including waterfowl), small mammals, and insects today continue to become mired in surface asphalt emanating from the old excavation sites. Inadvertent entrapment could have occurred when the surface of the asphalt was concealed beneath the surface of streams or pools, or covered by windblown dust and leaves, or through lack of experience or attention on the part of the unfortunate captive. It seems certain that fluvial sedimentary processes were also involved in burying animal or plant remains trapped in the asphalt and hence ultimately responsible for their fossilization. However, at least some of the fossil remains, both vertebrate and invertebrate, were originally the result of normal sedimentary accumulation (Akersten, 1991) and some may have been transported for considerable distances; such fluviatile accumulations were secondarily impregnated by asphalt (Doyen and Miller, 1980; S. E. Miller, 1983; Scott, 1989).

The former storage location of the La Brea fossils in the basement of the Natural History Museum of Los Angeles County provided interesting confirmation of the entrapment potential of asphalt. When all the Rancho La Brea fossils were finally transferred to the Page Museum during the mid-1980s, a number of recently mummified rodents and cockroaches were found trapped by asphalt which had leaked from the buckets of unprepared bones collected during construction of the Page Museum in 1975.

NATURE AND PRESERVATION OF THE FOSSIL REMAINS

The peculiar character of the embedding material—a heavy oil or soft asphalt in some instances but frequently a granular asphalt—has been largely responsible for the excellent preservation of the animal and plant remains. To be sure, the softer animal tissues have disappeared, leaving only the harder parts such as skulls, teeth, and bones, for identification of the original animal. However, fossil occurrences that are rarely encountered elsewhere—

Figure 9. Typical excavation at Rancho La Brea showing exposure of skulls and bones of Pleistocene animals in the asphalt. Note skull of wolf near top, jaws of bison at middle, and hip bone of large ground sloth at bottom. Photograph by John C. Merriam.

tracheal rings of birds, the fleshy parts of leaves, iridescent coloration in beetle wing cases—are commonly preserved at Rancho La Brea (Shaw and Quinn, 1986). The unusual preservation of Rancho La Brea fossils extends to aspects of their biochemistry (Akersten et al., 1983). Up to 80 percent of the original collagen remains in the bone (Ho, 1965) and even its microstructure is often well preserved (Doberenz and Wyckoff, 1967). Amino acid ratios in this fossil collagen were used by Ho (1967) to demonstrate that the normal body temperatures of extinct Rancho La Brea mammals were similar to those of their closest living relatives.

Specimens of mammals and birds are particularly well represented, the various skeletal elements and skulls of these types frequently forming thickly matted accumulations as shown in Figure 9. The distribution of these deposits is usually irregular, the masses of bones occurring as pocketlike concentrations in the asphalt. Merriam noted that in a mass comprising less than four cubic yards, a careful count indicated the presence of more than

50 heads of dire wolf, at least 30 skulls of the saber-toothed cat, and numerous bones of bison, horse, sloth, coyote, birds, and other forms. Reptiles and amphibians are only sparsely represented. Insect remains occur, and are common in some deposits. The record of the plants includes a great variety of species and the larger material consists principally of pieces of wood, although pollen, leaves, cones, and seeds have been recovered.

Skulls, teeth, and skeletal elements found at Rancho La Brea have come down through time practically unchanged from their original state. During the period of entombment the investing substances thoroughly impregnated the bones and teeth and may accumulate in quantity in the remote sinuses of skulls or in the marrow cavities of the long bones. Save for a prevailing black or brown color imparted to the bone, little difference in state of preservation is noted between this material and that of modern forms. The brain and nasal cavities of skulls are filled with asphalt which frequently has carried into these chambers the skeletal remains of small mammals and birds. The soft matrix has likewise held intact the largest as well as some of the smallest bones. It is interesting to note that from some mammalian skulls have been recovered the tiny bones of the inner ear. Teeth are generally well preserved and often retain the evidences of wear to which they were subjected in life (see Figure 10). Some herbivore teeth even contain fossilized plant fragments that document what the animal had eaten before it died. Limb elements exhibit not only the form and manner of articulation with adjacent bones but also the courses of nerves and blood vessels and the place of attachment of important tendons and ligaments. Injured and diseased bones and teeth occur in the collections. Fractured bones that had healed in life are found among both the mammalian and avian remains (Figure 11). Bone lesions due to pathological disturbances other than those arising from fractures are not uncommon. Materials displaying the characters of wear and disease emphasize that the organic remains from the asphalt were parts of once living creatures.

Curious as it may seem, the epidermal structures of vertebrates are not preserved. Thus, no preservation has been noted of hair or feathers, of the strong, horny nails or claws in mammals, or of the horny beaks and talons in birds. On the other hand, parts of the chitinous bodies of insects are present in the asphalt. Wood found at Rancho La Brea has a remarkably fresh appearance and, as may be expected, burns readily. Cones of pine and cypress have been thoroughly impregnated by the oil and exhibit their structures in considerable detail. Occasionally leaves are found, but more frequently only the impressions remain, in which, however, the details of the venation can still be discerned.

Parts of an individual skeleton are sometimes associated in the asphalt, although movement in the mass has tended to separate the bones laterally and vertically. It appears probable that the remains of any single skeleton are to be found in relatively close proximity. However, with the exposure of bodies at the surface of an asphalt trap it seems fair to assume that some carcasses were dismembered, and that some parts were strewn about on the ground adjacent to the borders of an active seep and were damaged, destroyed, or devoured before submergence in the asphalt preserved the remainder. Subsequent movement of bones within the asphaltic material is indicated by what has been termed "pit wear." The larger bones occasionally exhibit grooves or cuts which apparently cannot be ascribed to the work of predatory beasts. The abrasions may be deep and in some specimens have nearly sheared an individual bone in two. Specimens are known in which the outer surfaces are almost entirely destroyed by such wear.

In contrast to these specimens, other skeletal elements in the collections exhibit surface effects clearly due to attrition by scavenging organisms, presumably carnivores and rodents (Figure 12). Relatively large abrasions have been noted in which chips of bone two or more centimeters in length have been flaked off or broken away. In some cases the bite has been strong enough to expose the marrow cavity. Small abrasions also occur, usually

Figure 10. Superior view of right lower jaw of the large dire wolf (*Canis dirus* (Leidy)). Specimen belonged to an old animal in which the teeth were considerably worn during the life of the individual. LACMHC 2301-R-172, Page Museum collection; Rancho La Brea Pleistocene.

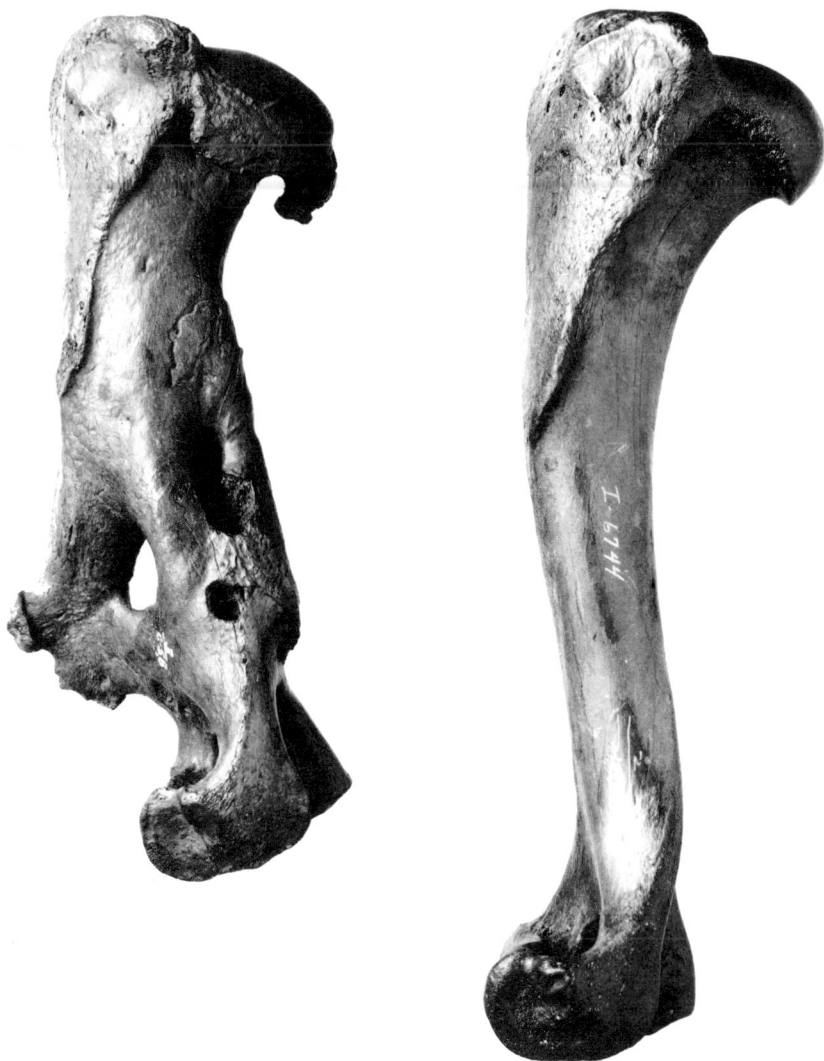

Figure 11. Lateral views of two specimens of left upper arm bone of dire wolf *(Canis dirus* (Leidy)). Figure on left (LACMHC 7122), specimen showing a healed oblique fracture with an abnormal bone growth; figure on right (LACMHC 16744), a normal specimen of the same bone. Both to same scale. Page Museum collection; Rancho La Brea Pleistocene.

in the form of grooves approximately 1.5 mm (1/16 in) in width, and sometimes parallel. An individual groove may show on closer inspection minute transverse ridges representing stages in the production of the groove by the chisel-like edge of the incisor teeth of rodents. Occasionally the two types of marking are superimposed. It is apparent also that the mammals intent upon breaking or gnawing a particular bone found a convenient grasp along the more pronounced borders, for the latter are often scarred. Skeletal remains exhibiting these features may have furnished a source of food coveted particularly by the strong-jawed carnivores and by the smaller gnawing forms.

The exposure of mammalian materials for any length of time at the surface of an asphalt seep, or in its immediate vicinity, is indicated not only by the markings left by other

Figure 12. Two views of right tibia or shin bone of the American lion (*Panthera leo atrox* (Leidy)) showing tooth marks left by carnivores and rodents. Figure on left, posterior view; figure on right, lateral view. LACMHC 2908-R-5, Page Museum collection; Rancho La Brea Pleistocene.

mammals, but also by the type of preservation of the compact bony tissue of the skull and skeletal elements. Individual bones usually retain their smooth external appearance. Some specimens, however, exhibit quite striking effects of weathering, and although now thoroughly penetrated and stained by the asphalt, they appear so closely similar to weathered skeletal remains found lying on the ground surface at the present time as to indicate similar weathering before these objects were finally covered by the asphalt.

Fossil insect remains have been found in the asphaltic fillings of cavities of skulls and limb bones belonging to extinct animals. An interesting example is that of a broken end of an upper arm bone of the great condor-like *Teratornis* in which were found, nestled among the cellular parts of the walls bounding the pneumatic (marrow) cavity, a number of puparia of a blowfly. The fly responsible for these larvae or larval cases, obviously deposited while the bird bone was still comparatively "fresh," was related to the black blowfly and to the screw-worm fly. As one of the ubiquitous organisms to be expected in the presence of death and during putrescence, this bit of evidence is graphically revealing. As a matter of fact, certain insects preserved in the asphalt are near relatives of kinds that today are present and characteristic of individual stages which mark the long cycle of disintegration that a carcass undergoes. It mutely demonstrates that ultimate entombment of remains at Rancho La Brea was at times a comparatively slow process (see page 16). It also points to the existence, perhaps prevalence, of offensive odors in and about the asphalt traps during their active period.

MAMMALS

HUMAN REMAINS

Portions of a human skull (Figure 13) and associated skeletal remains were encountered in the course of the Natural History Museum excavations in Pit 10 at a depth extending from approximately 2 to 3 m (6 to 10 ft). The remains occurred in one of two pipes or chimneys connecting from an asphaltic reservoir below with a surface flow above. The material filling the pipe consisted of a viscous mass containing sand and hardened lumps of weathered asphalt. Presumably, the material was derived in part from below and in part from above. The human remains belonged to one individual and are clearly those of our modern species. Judging from the structural features of the skull, this individual was not unlike those aboriginal people who lived on the Channel Islands and in the coastal province of southern California prior to the arrival of European explorers. Merriam (1914) concluded from a preliminary study of the human find that "the age of this specimen may perhaps be measured in thousands of years, but probably not in tens of thousands."

Possible association of man with the extinct condor-sized teratorn (*Teratornis*) and with extinct Pleistocene mammals at Rancho La Brea gave the occurrence special potential significance. However, the bird and mammal assemblages found in association with the human remains in Pit 10 suggest a lesser antiquity. Howard and Miller (1939) concluded from a study of the birds that: "Looking at the avifauna of Pit 10 as a whole, we find here an important link between the typical Pleistocene and the typical Recent. Nine of the sixteen extinct species associated with Pleistocene Rancho La Brea are present in Pit 10, but all are so reduced in numbers as to be only one-tenth as abundant as they were in the Pleistocene. We see in this assemblage, then, an intermediate stage in the extinction of birds usually considered to be typically Pleistocene. The avifauna as a whole, however, resembles that of the Recent more closely than that of the Pleistocene. In terms of overall age, therefore, Pit 10, in which the *Homo* remains were found, should fall into the Recent."

Similarly, the associated mammals are for the most part types more characteristic of the present rather than of the Pleistocene. There are no indications of the presence of the formidable carnivores so well represented in the typical Pleistocene asphalt at Rancho La Brea, and among the herbivores only a horse is recorded. It has not yet been shown that this species was identical with the Pleistocene *Equus occidentalis*.

Figure 13.   Skull of La Brea Woman (LACMHC 1323) discovered during the excavation of Pit 10 in 1914.

The skeleton of "La Brea Woman" has been radiometrically dated at 9,000 ± 80 BP (Berger et al., 1971), confirming the estimates by Merriam (1914), Howard and Miller (1939), and Stock, and postdating the bulk of the Rancho La Brea biota which accumulated between 11,000 and 34,000 years BP. A critical reexamination of the human material by Bromage and Shermis (1981) confirmed that these remains represent a female that stood about 144 cm (4 ft 8 in) tall. On the basis of epiphyseal union and dental evidence, they also estimated the age of the specimen to be between 25 and 30 years old at the time of death and concluded that the morphology of the pre-auricular groove in the pelvis indicated that La Brea Woman had borne at least one child. More recently, Kennedy (1989) estimated the age at death to be 17 to 18 years. Scott (1991b) has suggested that pelvic evidence of childbirth in La Brea Woman is inconclusive.

Early workers (e.g., Hrdlicka, 1918; Heizer, 1943) explained the presence of human remains in the Rancho La Brea asphalt deposits as the result of accidental entrapment. Bromage and Shermis (1981) interpreted cranial fractures in this specimen as traumatic lesions, suggested they were caused by blows from a grinding stone found 10 cm (4 in) from the skeletal remains, and hypothesized that La Brea Woman was Los Angeles' first documented homicide. Reynolds (1985) interpreted the broken grinding stone to have been ceremonially defaced and, on the basis of association of shell and stone artifacts and the skull of a small domestic dog with the partially complete human skeleton, suggested that this occurrence constituted secondary reburial rather than primary interment, accidental entrapment, or the concealment of a homicide victim.

The presence of *Homo sapiens* at Rancho La Brea is not limited to the evidence encountered in Pit 10, for scattered materials of human origin have been uncovered in some of the sites from which the more ancient brea fauna has come. This is particularly true for Pits 61 and 67, whence have come a wooden bunt foreshaft for an atlatl dart and three broken atlatl dart foreshafts. These specimens are recorded as coming from a depth extending from 2.4 to 5.5 m (8 to 18 ft) beneath the surface. Woodward (1937) called attention to the fact that these specimens are heavier and of cruder workmanship than those discovered by Harrington in Gypsum Cave, Nevada. Woodward stated further that the presence of the foreshafts in the Rancho La Brea deposits indicates that an atlatl-using people once inhabited or at least penetrated into Southern California and possibly were contemporaneous with animals now extinct but which are found in the brea.

Human artifacts from the asphalt deposits were later briefly discussed by Marcus and Berger (1984). One of the three broken atlatl dart foreshafts from Pits 61 and 67 has been dated at 4,450 BP (Hubbs et al., 1960). A cogged stone from Pit 67 is representative of a cultural industry present in southern California from 5,500 to 1,000 BP (Salls, 1980). Worked antler, bone, wood, and shell were also recovered from that site (Woodward, 1937). *Smilodon* bones (Pits 4 and 77) and bone of bison (Pit 4) and American lion (Pit 3) bear surface features that G. J. Miller (1969) interpreted as cut marks of possible human origin, although this interpretation was discounted by Jefferson (1988). Marcus and Berger (1984) suggest that some of the marine mollusks recovered from younger asphaltic horizons could have been transported by human agency, or introduced from below via an asphaltic vent, or represent contamination from other nearby excavations. Most objects accepted as artifactual are from deposits with mixed stratigraphy and provide no clear-cut evidence for humans at Rancho La Brea prior to 9,000 BP (Marcus and Berger, 1984).

GROUP REPRESENTATION AMONG THE MAMMALS

The most striking feature of the Rancho La Brea mammalian assemblage is the preponderant number of predatory forms. In this aspect the fauna differs noticeably from existing or most other extinct assemblages throughout the world. This unique character of the Rancho La Brea fauna appears to result from the fact that victims mired in the asphalt traps provided a lure that was particularly effective in attracting flesh-eating mammals to such seeps.

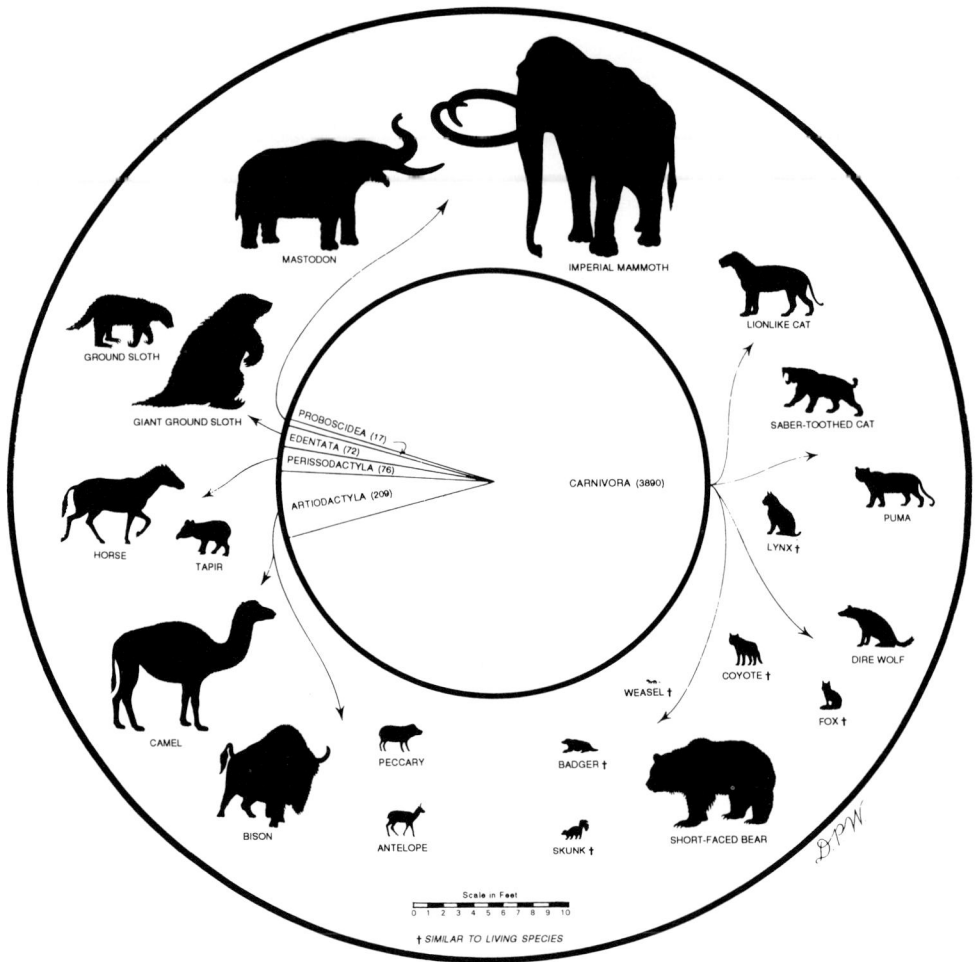

Figure 14. Diagram illustrating relative number of individuals in the mammalian orders (except rodents, lagomorphs, insectivores, and bats) occurring in the Rancho La Brea Pleistocene fauna. Note the preponderance of predatory forms.

A census of the Pleistocene mammals represented in the collection of the Page Museum, reveals a total of more than 10,000 individuals. By far the largest number of animals comprising this population are carnivores. In contrast, the herbivores, or plant-feeders, are distinctly fewer in number. Figure 14 graphically illustrates the relative proportions of the various orders of Rancho La Brea mammals. Among the Carnivora the largest number of individuals is included in the canid (or dog) family, with the felids (or cats) forming the next largest group. The dogs constitute approximately 57 percent of the carnivore population, the cats 32 percent. Each of these families greatly exceeds in number the bears and mustelids. Among the plant-feeders, the family having the greatest number of individuals is the Bovidae (bison). Then follow in turn the horses (Equidae), mylodont ground sloths (Mylodontidae), camels (Camelidae), antelopes (Antilocapridae), mastodonts (Mastodontidae), mammoths (Elephantidae), and deer (Cervidae). Last come the peccaries (Tayassuidae) and tapirs (Tapiridae). Precise figures are not currently available for most of these groups.

Earlier excavations concentrated on the larger and more spectacular extinct mammals and largely ignored smaller elements of the fauna, but the number and diversity of rodents and other small mammals from the asphalt deposits has increased dramatically as a result of the excavation of Pit 91 during the past 20 years.

In those mammalian families which include extinct forms as well as species that have persisted from the Pleistocene into Recent time, the extinct forms are always represented by a greater number of individuals. In other words, the typical Pleistocene species in the Rancho La Brea fauna are relatively more important elements in this assemblage than those types which are characteristic of the Recent epoch but whose range extends also into the Pleistocene. Thus, within the dog family the number of dire wolves is far in excess of that of the coyotes and gray wolves. The modern gray fox is known by only a few individuals. Among the cats, the sabertooth, a creature now extinct, is much better represented in the fauna than the great lionlike feline. The latter, while specifically different, is still closely related to some of the large living cats. Both the sabertooth and the lionlike cat greatly exceed in numbers the puma, jaguar, and lynx. Among the ursids, the short-faced bears are twice as abundant as individuals of black bear or grizzly bear. Among those hoofed mammals that permit a similar comparison the same relationship is evident. Within the antilocaprid family, for example, an extinct form (*Capromeryx minor* Taylor) far outnumbers fossilized representatives of the living pronghorn.

The relative abundance of those Pleistocene mammals that became extinct before the present day provides an index of the antiquity of the assemblage. In other words, the position of the Rancho La Brea fauna in geologic time is suggested by the presence of many typical Pleistocene mammals. Were this fauna actually of the Recent epoch rather than of the Pleistocene, its composition would unquestionably include a lesser number of typical Pleistocene mammals and a greater proportion of those forms that survive today. The Recent fauna is an outgrowth of that of the Pleistocene. The lack of great diversity of mammals in North America at the present time is largely but not entirely, due to the extinction of many of the older forms. Exact information as to time of disappearance of these earlier types is of great importance in determining the position of Rancho La Brea in the chronological events of the later Pleistocene, but this is discussed elsewhere (see pages 77–78).

Other factors that exert an influence on the presence or absence of different kinds of mammals are found in environmental conditions and in the habits of the animals themselves. The extraordinary numbers of dire wolves and sabertooths reflect the conditions which prevailed in the immediate vicinity of the asphalt accumulations and the availability of food. The dire, or grim, wolves undoubtedly preyed upon the large, cumbersome, and slow-moving mammals with which they were associated. It appears probable that the dire wolves occupied a niche in the North American animal world of the Pleistocene comparable to that held by the hyenas in the wildlife of Africa and Asia today. The asphalt traps, with their living hosts and partly devoured, dismembered carcasses, doubtless offered very suitable feeding grounds for both predators and scavengers.

The sabertooth, in contrast to the great lionlike cat, was certainly not a predatory form that depended on fleet-footedness for hunting its prey. This is clearly indicated by the fact that the lower segments of its limbs (the two bones of the forearm and those of the foreleg) are distinctly shorter than those of a typical running form like the lion. The sabertooth instead exhibits an organization admirably adjusted to grappling and fighting at close quarters. Truculent to the extreme, this creature found its victims among the slow-moving mammals and the stationary live bait of the asphalt traps. The unfair advantage which the saber-toothed cat possessed around the borders of a brea seep was apparently compensated only by its own high mortality.

That the country surrounding the asphalt deposits was essentially similar to that of the Los Angeles area in historic times is suggested both by the mammalian assemblage as a whole and by the geologic history of the region. The climate may have been as equable as it is today, with only slightly greater precipitation and somewhat lower temperature. A plain, or open rolling country on which grew an interior, semiarid type of vegetation, and where grass- and herb-covered surfaces were interspersed with copses of trees and brush, seems to have favored the existence of a diverse population of hoofed mammals. In this environment, forms such as the bison, horse, mylodont ground sloths, mammoths, camels, and antelopes would normally be found. Associated with these herbivores were typical cursorial carnivores like the lionlike cat and the coyote. Other important members of the community were, of course, the dire wolves and the sabertooths. One might regard this assemblage as the resident population.

In contrast to this group are those forms which did not live habitually in the region, but which occasionally penetrated it from adjacent areas. This assemblage includes species that are only sparsely represented at Rancho La Brea, among them the megalonychid ground sloths with browsing habits, timber wolves, peccaries, deer, mammoths and mastodonts, and tapirs. The presence of these animals perhaps may be accounted for by intervals when the climate was slightly more humid than usual.

Stock (1930) interpreted the region to have been sufficiently wooded or brush-covered to afford shelter to a number of diminutive pronghorn antelopes (*Capromeryx*). He suggested these animals may have occupied a niche in the environment similar to that of the African dik-dik and duiker antelopes, emerging from cover at dusk to feed. However, given the open-country adaptations of their larger extant (*Antilocapra*) relatives, it is not implausible that *Capromeryx* may have behaved more like the small Thomson's gazelles of the African grasslands.

It is a curious fact that few members of the raccoon family are recorded among the Pleistocene mammals found in the brea beds. Two members of this raccoon family—the ringtail (*Bassariscus astutus* (Lichtenstein)) and the raccoon itself (*Procyon lotor* Linnaeus)—are now known by single specimens from Pit 91 (Akersten et al., 1979). The rarity of these mammals may be due to their nocturnal habits, as the adhesive property of the asphalt becomes noticeably less effective in the evening when it is cool.

Small ponds and streams were present in the immediate vicinity of the asphalt beds during Pleistocene time. These may have served both as attractions and traps. Animals coming to the region to drink may have waded into the water to become mired in the asphalt beneath. Occasionally, liquid asphalt may have been mistaken for water and certainly some sites have yielded a considerable number of fossilized water birds. However, the environment at Rancho La Brea during the Ice Age was probably not strictly comparable to that of a water hole on the African plains today. The large number of trapped animals does not imply that the surrounding area was as devoid of water as are the drier parts of the African savanna at the present time. On the contrary, streams such as the predecessor of the present Los Angeles River were present at no great distance away. These, even during the driest seasons, would have had a sufficient flow of water to sustain the animal life of the region.

Moreover, interpretation of faunal variety displayed by the trapped assemblages should take into account seasonal and diurnal temperature fluctuations that would have affected the viscosity and adhesiveness of the asphalt. Because extruded asphalt only becomes sticky when warm, nocturnal animals, or those present only during the cooler seasons, are less likely to have been trapped in asphalt than animals present and active during the warmer parts of the year.

INSECTIVORA (Shrews)

Only two species of shrews have thus far been recognized in the Rancho La Brea collections. A few bones have been referred to the ornate shrew (*Sorex* cf. *S. ornatus* C. H. Merriam), which is the only long-tailed shrew now living in the region. Remains of the desert shrew (*Notiosorex crawfordi* Coues) are fairly abundant, outnumbering the ornate shrew ten to one. Regarding this form, Compton (1937) remarked: "The occurrence of this shrew at Rancho La Brea offers an interesting comparison of its distribution during Pleistocene times with that of today. At present this species is restricted to the southwestern states, and is considered rare. It would appear, then, that *N. crawfordi* either ranged farther north and west along the Pacific slope during the Pleistocene, or that our knowledge of its modern distribution is incomplete."

As a result of recent excavation in Pit 91, the broad-handed mole *Scapanus latimus* is also represented in the Rancho La Brea fauna by a single right humerus (Akersten et al., 1979).

CHIROPTERA (Bats)

Chiropterans are rare in any fossil assemblage, in part because of their fragility and small size, but the hoary bat (*Lasiurus cinereus* Palisot de Deauvois) has now been recovered from the Pit 91 excavation (Akersten et al., 1979) and the pallid bat (*Antrozous pallidus* (le Conte)) has also been identified in previously collected material. The pallid bat feeds near the ground, often landing to pick up beetles, crickets, and other large insects (Kurtén and Anderson, 1980), which may account for its presence in the asphalt deposits. The hoary bat is generally solitary and roosts in the foliage of trees rather than in caves.

CARNIVORA

CANIDAE (Wolves, Coyotes, Foxes)

Individuals of the canid, or dog, family are the most commonly occurring mammals in the Rancho La Brea assemblage. No other group of carnivores is represented by so great a number of individuals, although the cats make a close approach in this respect. The dogs from the asphalt deposits were first described in detail by Merriam (1912b).

An unusual feature of this group is the large representation of the dire, or grim, wolves *Canis dirus* (Leidy). These forms were presumably very widespread over the North American continent during Pleistocene time, for their remains have been encountered at a number of fossil localities. Originally described from Pleistocene deposits in the Mississippi Valley, the dire wolves have since been recognized as far east as Florida and as far south as southern Nuevo León and the Valley of Mexico. In addition to the occurrence at Rancho La Brea, records of their presence in California have been found in Pleistocene deposits of Livermore Valley, in beds of similar age along the border of the San Joaquin Valley, at McKittrick and Carpinteria, and at San Pedro. Specimens from California and Mexico were smaller and had shorter limbs than those encountered in the central and eastern parts of the United States (Kurtén and Anderson, 1980).

Canis dirus, shown in Figure 15, is a large species of wolf but was approximately 8 percent smaller than the largest known representatives of the northern timber wolf living today in northern Alberta, Canada. *Canis dirus* was, however, larger than the timber wolves that are found today at more southerly latitudes in North America. In either case, the external appearance and habits of the extinct and living species must have been quite different.

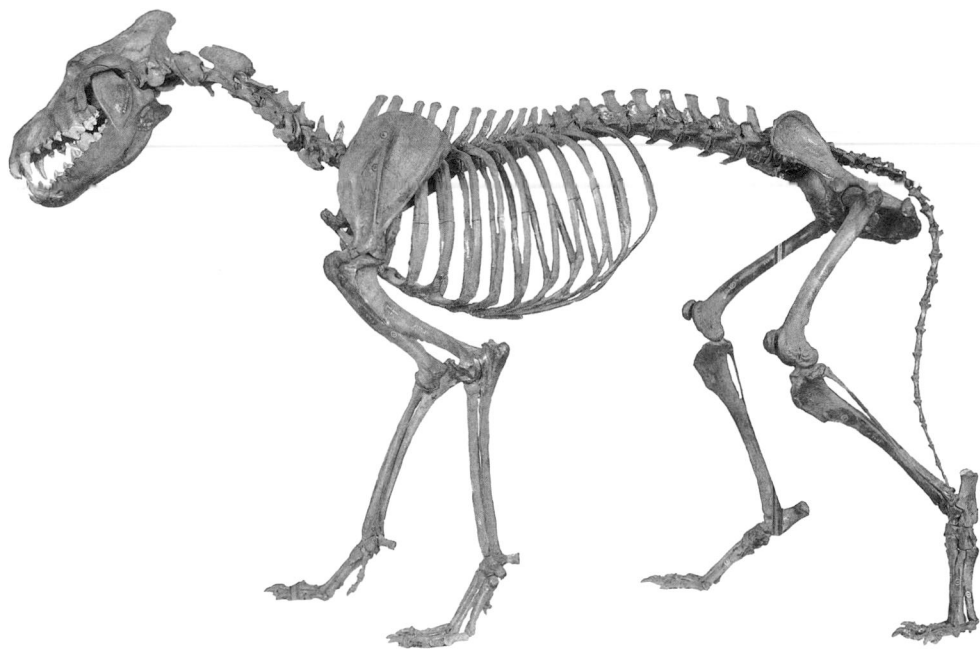

Figure 15. Composite skeleton of large dire wolf (*Canis dirus* (Leidy)). Page Museum collection; Rancho La Brea Pleistocene.

Canis dirus had a large and heavy head, a relatively small brain, a massive dentition, as well as large shoulder blades and pelvis. Their powerful jaws and teeth would have been advantageous in pulling down large prey and furnished a powerful biting mechanism capable of crushing large bones. It appears not unlikely that the dire wolf occasionally resorted to carrion for food.

Differences in appearance must have existed between the dire wolf and the timber wolf as a result of the proportions of front and hind limbs and of particular limb segments. In the extinct wolf, the foreleg is shorter than the hind leg, albeit the difference is small. In addition, the lower segments of the limbs in the dire wolf, particularly in the hind limb, are shorter relative to the length of the upper segment (upper arm bone and thigh bone) than in the timber wolf. These differences may signify that the dire wolf was not so fleet of foot as the timber wolf.

Concerning the habits of these creatures, Merriam (1912) has remarked: "The form of the skull suggests that the head was normally held low and was often used in hard pulling and hauling of heavy bodies. The great number of individuals of C. *dirus* found at Rancho La Brea suggests that the wolves of this species sometimes associated themselves in packs, and that groups of considerable size may have assembled to kill isolated ungulates and edentates. Particularly the young, aged, and injured, when they could be separated from their associates, would be the natural prey of the great wolf, but adults in normal strength may also have succumbed to the combined attack of several of these powerful animals."

As in the case of the saber-toothed cat, the collection of dire wolf bones includes a number of specimens showing fractures and abnormalities in bone growth. Luxations were probably due to injury (see Figure 11, page 18) and were followed by chronic infection in several instances. Some of the traumatic injuries observed in fossil specimens were the result of blows to the head and trampling of the forelimbs; similar injuries have been

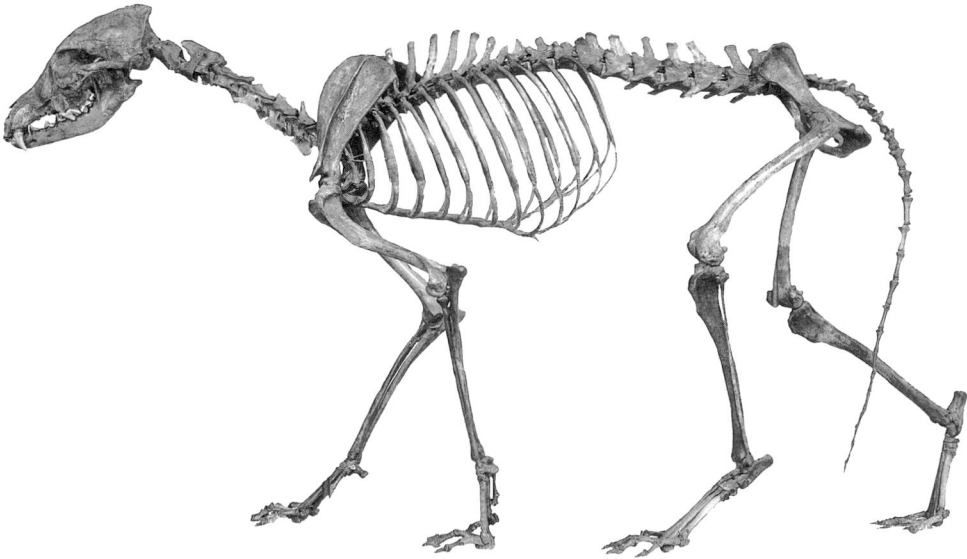

Figure 16. Composite skeleton of fossil coyote (*Canis latrans* Say). Page Museum collection; Rancho La Brea Pleistocene.

observed in living wolves and are incurred from large prey animals during hunting episodes. It is interesting that the type of injuries characteristically incurred by dire wolves were different from those typical of *Smilodon*, the other common large Rancho La Brea predator, and this suggests a totally different method of prey capture.

Dire wolves were undoubtedly the major predators of the Los Angeles Basin during the waning phases of the Pleistocene Ice Age. Like spotted hyenas, large wild canids (wolves, the Cape hunting dog, etc.) tend to run their prey down through cooperative sequential chasing by different members of the pack rather than by stalking and pouncing by individuals. It is thus possible that both dire wolves and their prey sometimes became inadvertently mired in surface asphalt through inattention to the terrain over which they were running.

In contrast to the hundreds of individuals of the dire wolf are the very limited number of canids related rather closely to the modern gray or timber wolves. This form, represented mainly by skull material, differs from the typical dire wolf in certain structural characters of the skull and dentition. Examples include specimens that were originally described as *Canis milleri* (Merriam) and *Canis furlongi* (Merriam) but that are now believed to be examples of the extant timber wolf (*Canis lupus* Linnaeus).

The coyotes rank next to the dire wolves and sabertooths in number of individuals found in the asphalt, although they are only one-eighth as numerous as the dire wolves. The coyotes are represented by one species (Figure 16), originally identified as *Canis orcutti* (Merriam) but now recognized as the extant species *Canis latrans* Say which still occurs today in the Los Angeles area. Pleistocene coyotes tended to be larger than their extant counterparts (Kurtén and Anderson, 1980), and the specimens previously identified as *Canis orcutti* or *C. andersoni* Merriam are now interpreted as individual variants of the extant coyote species.

The presence of many coyotes at Rancho La Brea is presumably due in large measure to the prevalence of their natural prey, namely, small mammals and birds held captive in the asphalt or hovering about the asphalt seeps. That coyote fossils are decidedly less

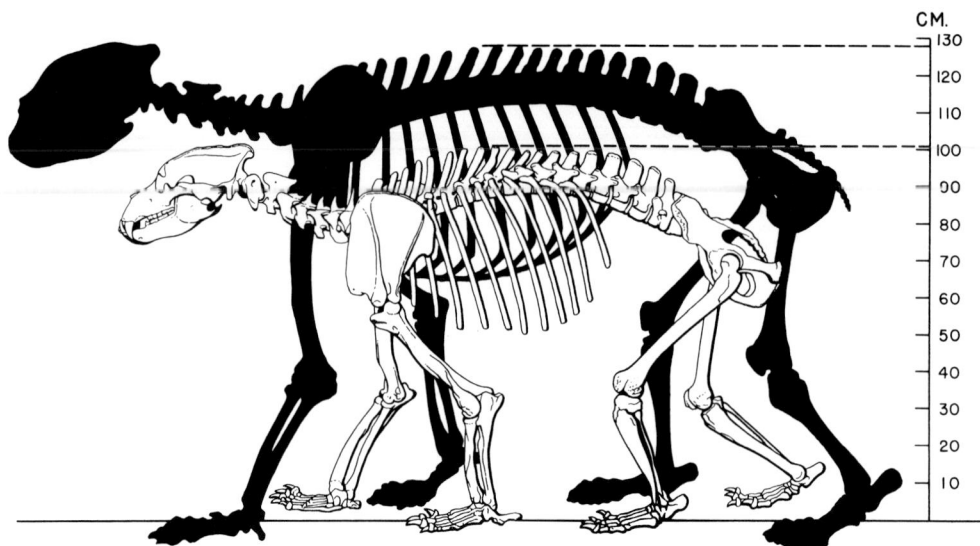

Figure 17. Comparison of the skeleton of the Pleistocene short-faced bear (*Arctodus simus* (Cope)) from Rancho La Brea in silhouette, with that of the Recent but extinct California grizzly (*Ursus arctos horribilis* Ord) in outline. Note the great difference in size.

common than those of dire wolves may reflect the actual proportion of these two carnivore species in the vicinity of Rancho La Brea during the period of accumulation of the asphalt beds. On the other hand, it may be due to the coyotes possessing greater intelligence or smaller body weight, either of which may have aided these creatures in avoiding the dangers of the asphaltic traps. With the passing of the Pleistocene and the extinction of the dire wolves, the coyotes established themselves as the most important group of carnivores in the region.

Several specimens of the domestic dog *Canis familiaris* Linnaeus have been recovered from the younger brea deposits. A large canid mandible recovered from Pit 61/67 was originally attributed to a separate species *C. petrolei* Stock (Reynolds, 1979). A skull and associated postcranial bones of a smaller individual were found with the remains of La Brea Woman and these have been interpreted as evidence of a human reburial (Reynolds, 1985).

A gray fox identical with the modern species (*Urocyon cinereoargenteus* Schreber) is also recorded. The desert kit fox (*Vulpes macrotis* Merriam) is absent, although it occurs in the asphalt at McKittrick.

URSIDAE (Bears)

Among the bears discovered in the fossil record of Rancho La Brea, three distinct types are recognized. Of special interest is the short-faced bear (*Arctodus simus* (Cope)). This form differs in a number of structural characters from the living bears of the North American continent (see Figure 17). As its name implies, the short-faced bear possessed a shortened face and somewhat crowded front cheek teeth in contrast to the black bear. Moreover, while the number of teeth in the upper and lower jaws is similar to that in existing bears of North America, the carnassial or principal cutting teeth (upper premolar 4 and lower molar 1) are somewhat more like those in typical carnivores, such as the dogs, than are the corresponding teeth of the grizzly or black bears. Doubtless, the short-faced bears were more carnivorous in their habits than were the true bears. Short-faced bears were characterized also by very large size, in which respect they resemble the great brown or Kodiak bears of the coastal region of Alaska. They were undoubtedly the largest flesh-eating mammals occurring at Rancho La Brea. More than 700 elements of *Arctodus simus* from the Rancho La Brea deposits represent a minimum of 30 individuals. Large and small adult forms from Rancho La Brea and elsewhere indicate this species was strongly sexually dimorphic (Cox, 1991).

The short-faced bears enjoyed an extensive distribution over the North American continent in Pleistocene time, remains of these creatures being found in the Yukon, in Pennsylvania, Kentucky, and Texas, and at a number of localities in California. They, or their close relatives, were also widely distributed in South America during the Pleistocene. This group of bears, while now entirely extinct, is more closely related to the spectacled bear of the Andes than to any living North American type.

The true bears are represented in the Rancho La Brea fauna by a black bear (*Ursus americanus* Pallas), the remains of which include an immature male skull, and by a grizzly (*Ursus arctos horribilis* (Ord)). These forms are closely related to types now living in the California region, or existing here during the historic past. The fossil black bear possessed relatively large grinding teeth, in which character it differs from its living relatives. Kurtén (1960) described a female skull of a grizzly bear from Pit 10. This species is known only from Holocene horizons at Rancho La Brea; elsewhere it replaced the short-faced bear over much of its range (Kurtén and Anderson, 1980).

Within the bear family it is interesting to note that those forms most closely related to types now living in California were distinctly outnumbered by bears of a kind now extinct. However, with the disappearance of the short-faced bears, the black and grizzly bears established themselves as the prevailing representatives of the family in California.

MUSTELIDAE (Skunks, Weasels, Badgers)

The smaller carnivorous mammals of the Pleistocene of Rancho La Brea, particularly those of the mustelid family, are, like the rodents, closely related to living members of the group. Pleistocene mustelids of the western part of the United States, however, tend to be somewhat larger than their modern representatives. It is not surprising to find recorded in the asphalt deposits the striped skunk (*Mephitis mephitis* Schreber), spotted skunk (*Spilogale putorius* (Linnaeus), weasel (*Mustela frenata* Lichtenstein), and badger (*Taxidea taxus* (Schreber)), in view of the habits of these animals at the present time. The weasel, represented by 53 skulls, is by far the most abundant member of this family in the asphalt deposits.

The predaceous skunks and weasels feed on small mammals and birds, whose occurrence in and about the asphalt traps undoubtedly accounts for the presence of their natural enemies. The badgers, with food habits somewhat like those of the skunks and weasels, are fossorial. It appears not improbable that in some instances these mammals were trapped in their burrows during the exudation or outpouring of the asphalt. On the other hand, badgers are known to move about considerably over the surface of the ground and may have floundered on occasion into the asphaltic material.

FELIDAE (Cats)

Machairodontinae (Saber-toothed Cats). The cat family as recorded in the asphaltic deposits includes representatives of both the saber-toothed and true cat groups. Perhaps the most unusual kind in type of specialization is the sabertooth (*Smilodon fatalis* (Leidy)).

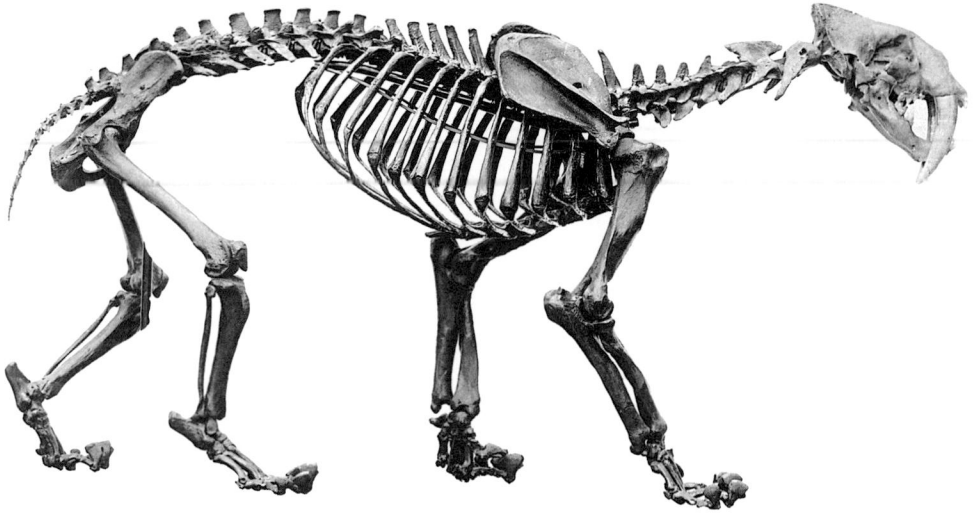

Figure 18. Composite skeleton of the saber-toothed cat *(Smilodon fatalis* (Leidy)). Page Museum collection; Rancho La Brea Pleistocene.

This form ranks next to the dire wolf in number of individuals found in the asphalt seeps and greatly outnumbers all other types of cats. G. J. Miller (1968) documented 2,100 *Smilodon* individuals from the Rancho La Brea collections at the Page Museum, based on crania and cranial fragments.

The sabertooth approximated in size the African lion, although the body and limbs were somewhat differently proportioned. In *Smilodon* the hind limbs are relatively light while the front limbs are strong and powerful (Figure 18). This sturdiness and strength is likewise shown by the rib cage and breastbone. The lower segments of the limbs are relatively short in comparison to those in the great extinct lion. *Smilodon* had limb proportions that, compared with modern felids, were most similar to those of the jaguar, *Panthera onca* Linnaeus (Gonyea, 1976a); both of these digitigrade species share features of the hind limb with plantigrade carnivores (Ginsberg, 1961). It seems probable that *Smilodon* used ambush and stalking techniques rather than rapid pursuit for the capture of prey (Gonyea, 1976b). A curious feature of this animal is the short tail, in which respect *Smilodon* exhibits a superficial resemblance to the lynx or bobcat.

Fundamental differences between the sabertooth and the lion or puma are perhaps most strikingly shown in the skull and dentition (Figure 19). *Smilodon* was long thought to have possessed a relatively small brain although Jerison (1973) demonstrated that the brain size of *Smilodon* differed only slightly from that of comparably large felids. The skull is curiously modified in adjustment to the great development of the daggerlike canine teeth in the upper jaw. Some of these modifications should be mentioned.

The external nasal opening has receded somewhat from its typical position in cat skulls. The hard palate develops rather prominent bony ridges which run the length of this surface. In the ear region the sabertooth skull exhibits a remarkable character in the growth of the mastoid, furnishing thereby a greater area for attachment of muscles exerting a strong downward pull on the head. Certain parts of the lower jaw, in contrast to those in the true cats, are weakly constructed. Judging from the development of structures to which important muscles were attached, the lower jaw swung through a wide angle when the mouth was opened in attack, and the biting strength of this element may have been correspondingly weakened.

Figure 19. Lateral view of skull of saber-toothed cat (*Smilodon fatalis* (Leidy)). LACMHC 2001-2 (cranium), LACMHC 2002-L&R-2 (mandible); Page Museum collection; Rancho La Brea Pleistocene. After Merriam and Stock.

The dental battery of the sabertooth presents some rather unusual specializations. In this cat the dentition of an adult individual usually consists of 26 teeth, while in the lion or puma 30 teeth are present. In other words, compared to the true cats the sabertooth has lost a front premolar tooth on each side of the upper and lower jaw. The upper canines are great daggerlike teeth, considerably elongated in their long curvature and flattened transversely. The front and back edges of the crown of the canine are minutely serrated. *Smilodon* literally means saber tooth, a name well applied to this animal. The lower canine teeth are reduced in size and resemble in this character and in shape the lower incisors. In the cheek tooth region the principal cutting teeth have their blades compressed transversely and lengthened in fore and aft line in adjustment to a cutting or slashing action.

In attacking a large mammal like a mammoth, mastodont, or ground sloth, the sabertoothed cat would probably have sought a vulnerable spot on the body of its prey, gripped the victim with its powerful front limbs and claws and repeatedly stabbed with the upper canines, thus inflicting a jagged wound. In such an attack the lower jaw was capable of swinging downward giving a considerable gape to the mouth, the powerful head and neck muscles furnishing at the same time a strong thrust which accompanied the stabbing action of the teeth. The backward position of the nasal opening presumably permitted the animal to breathe with head plunged deeply into the side of its victim. The presence of a strongly corrugated gum covering the ridges of the hard palate may have rendered service in blood-sucking, as Stock and others suggested, but is more likely to represent strengthening of the cranium to counter stresses incurred during canine use (C. A. Shaw, pers. comm.).

Akersten (1985) postulated that the saberlike canines of *Smilodon* were used primarily in biting open the abdomen of large prey animals, arguing that the relatively fragile nature of the long and slender canines would not be effective for use in regions of the body where bone would be encountered immediately after penetrating the skin. A recent comparison of dental microwear patterns of *Smilodon* from Rancho La Brea with those of eight different extant carnivores suggested that *Smilodon* rarely ate bones but occasionally participated in scavenging activity (Van Valkenburgh et al., 1991).

A curious feature noted in an examination of the great collection of skeletal remains of the sabertooth is the relatively high frequency of lesions in particular elements. While pathological disturbances of the normal structure are occasionally noted in the skull and dentition, they are particularly evident in the limb bones and in the lumbar region of the vertebral column. Moreover, while fractures which have healed during the life of the individual are to be found in a number of bones of mammals and birds from the asphalt, abnormalities in bone growth were apparently rather prevalent among the saber-toothed cats. According to Dr. R. L. Moodie, who gave considerable attention to the study of the diseases affecting Rancho La Brea mammals and birds, most of the pathological conditions are to be attributed to injury with subsequent infection. Furthermore, cases of luxation and arthritis have been recognized. It is perhaps not surprising to find disturbances of the normal bone growth in creatures as savage as the sabertooth and dire wolf. Injuries were doubtless inflicted frequently in the combats that transpired when large numbers of these beasts gathered about the asphalt seeps.

Less evident are the causes which have contributed to the occurrence of abnormalities in the vertebral column of *Smilodon*. In the lumbar region, particularly, two vertebrae or as many as four may fuse to form a more or less solid tube due to the development of excess bony tissue along the sides and bottom of the vertebral series (see Figure 20). This malady resembles a pathological state occurring in humans in which a progressive ossification or formation of bone takes place in the muscle tissue lying adjacent to the lumbar vertebrae. Whether or not the peculiar habits of these creatures were in a measure responsible for this unusual fusion has not been determined, but the condition remains one that is strikingly characteristic of this group of predatory beasts.

Pathological injuries that occurred most frequently in the neck, chest and lumbar region of *Smilodon* (Heald, 1986, 1989; Shaw, 1989; Shaw et al., 1991) seem consistent with trauma incurred during catching prey animals and wrestling them to the ground. Septic lesions consistent with bites or other open wounds, and other debilitating traumatic injuries are also evident in the Rancho La Brea sample. The survival of individuals long after the receipt of crippling wounds has been interpreted to suggest cooperative social behavior in this species (Heald, 1989; Shaw et al., 1991).

Occasionally, skulls of the sabertooth are encountered in the asphalt in which one or both of the saberlike teeth were broken in life, an injury sustained apparently during feeding. In such specimens the broken edge of the saber exhibits a worn and smooth surface, clearly denoting the fact that the cat subjected the tooth to use after injury occurred.

Figure 20. A series of four anterior lumbar vertebrae of the saber-toothed cat *Smilodon fatalis* (Leidy). Upper figure, fused vertebrae II through V with ossification of the lateral muscle mass (LACMHC 277); lower figure, a series of corresponding vertebrae in which the segments have not coalesced. Page Museum collection; Rancho La Brea Pleistocene.

When the canine was broken during life, the cheek teeth also frequently show excessive wear, suggesting that the animal found greater need for these teeth after loss of the canines. Quite obviously, such individuals were at a decided disadvantage in their struggles for existence. However, an interesting finding by Van Valkenburgh et al. (1991) was that, despite their relative fragility, *Smilodon* canines were broken less frequently than those of living carnivores.

The recovery of a complete and semiarticulated *Smilodon* skeleton from the deposit discovered during construction of the Page Museum indicates that the metapodials and phalanges became progressively smaller from the second to the fifth digit (Cox and Jefferson, 1988), unlike most previous restorations based on incomplete or unassociated skeletal material. All ontogenetic stages of *Smilodon* except that of very young individuals have been recovered from the Rancho La Brea localities (Tejada-Flores and Shaw, 1984), the absence of the most immature forms perhaps indicating that denning areas were not located near the asphalt seeps.

The saber-toothed cats have had a long and eventful history in North America from the early Oligocene Epoch through to the late Pleistocene. Saber-toothed cats closely related to the Rancho La Brea species have been recorded in Florida, Nebraska, Texas, and Mexico, and similar forms are known from the Pleistocene of South America and western Europe.

Berta (1985) reduced the number of *Smilodon* species to two—*Smilodon gracilis* (Cope) from the late Pliocene to middle Pleistocene of North America and *Smilodon populator* Lund (including *Smilodon californicus* Bovard, *S. floridanus* Adams, and *S. fatalis*) from the middle to late Pleistocene of both North and South America, but her interpretation has not yet been universally accepted. We here follow Churcher's (1984) taxonomic interpretation of North and South American species. The subspecies *S. fatalis brevipes* was proposed by Merriam and Stock (1932) for some unusually short machairodontid metapodials from Rancho La Brea; whether this taxon represented a distinctive (sub)species of *Smilodon* or incursion of *S. populator* (characterized by shortened distal limb segments (Kurtén and Anderson, 1980)), has yet to be substantiated.

Homotherium serum (Cope), the scimitar cat, was a very rare component of the Rancho La Brea biota, being represented by a few metapodials and by canine teeth that are readily distinguishable in shape and size from those of *Smilodon*. Although it had a wide geographic range, *Homotherium serum* was never as abundant as *Smilodon* (Kurtén and Anderson, 1980) and probably became extinct between 18,000 and 20,000 BP (Graham, 1976b).

Felinae (True Cats). For the group of true cats, represented today by the African lion, Indian and Asiatic tiger, African leopard, South American jaguar, and all other types of cats, the asphalt deposits reveal a noteworthy record. Undoubtedly the most remarkable member of this group is the great American lionlike cat (*Panthera leo atrox* (Leidy)), male individuals of which were nearly one-fourth larger than any of the large living cats of Eurasia (Figure 21). While it has become customary to speak of this feline as a lion, the species has also been called a gigantic jaguar although, nothing is known regarding the pelage and external coloration of this animal. While differing in size from the large living felines the American lions are no less unusual because of their close structural similarity to modern species. In certain features of the skull, *Panthera leo atrox* is more like the African lion than the Indian tiger. There is a noticeable size difference between the sexes. Skulls of females from the brea approach in size the skulls of the larger male individuals of the South American jaguar. Without much question this great cat was the most formidable predatory mammal present in the Rancho La Brea assemblage. Only the short-faced bear exceeded it in size.

These powerful cats are not nearly so well represented in the asphalt as their cousins, the saber-toothed cats, although they are more common than the puma and lynx. Agile and strong of body and limb, fleet-footed, and doubtless possessing the superlative grace of line, surety of step, and stealth of approach so characteristic of felines, it is not difficult to perceive *Panthera leo atrox* as the greatest hunter of its time. Stalking prey in the open, depending on its great biting strength and speed in its attack on the larger herbivores, this magnificent creature was as characteristic of the North American continent during the Pleistocene as the lion is of the African savannas at the present time.

At least 80 individuals of the American lion have now been documented from the asphalt deposits, adult males outnumbering adult females by a ratio of three to two (Jefferson, 1991). Based on the dimensions of the femur, adult males were estimated to weigh 235 kg (517 lb) versus 175 kg (385 lb) for females. The American lion had a larger brain, relative to body size, than any of the Pleistocene or living lions of the Old World, and the high degree of cephalization suggests this form was gregarious and hunted in groups (Kurtén and Anderson, 1980). However, Jefferson (1991) interpreted the greater proportion of male *Panthera leo atrox* in the Rancho La Brea sample to be inconsistent with the pride structure typical of extant lions.

Specimens identical with the Rancho La Brea form were described many years ago from the Pleistocene of Natchez, Mississippi, and *Panthera leo atrox* has been recognized as

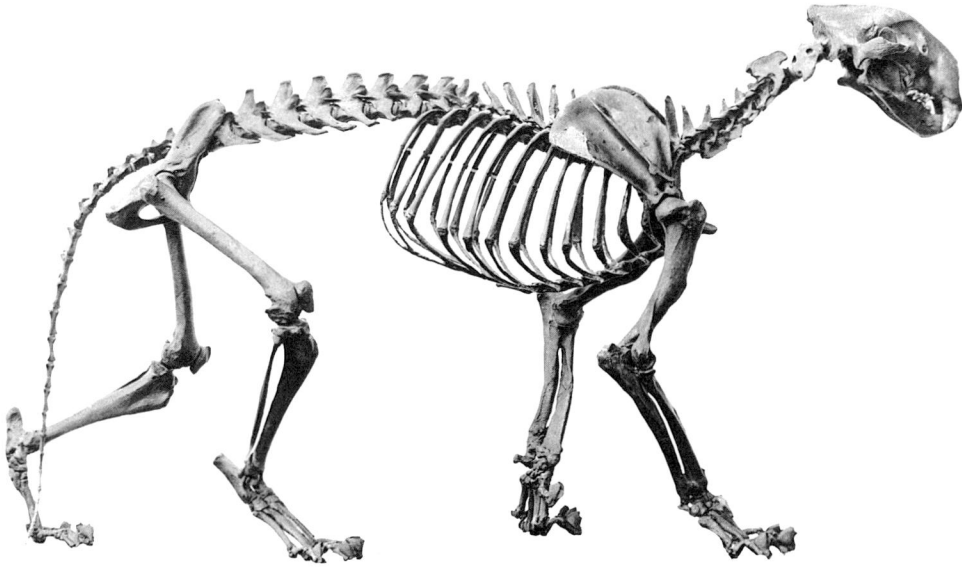

Figure 21. Composite skeleton of the American lion (*Panthera leo atrox* (Leidy)). Page Museum collection; Rancho La Brea Pleistocene.

far south as the Valley of Mexico and as far north as Alaska. During the glacial intervals, large cats closely related to the North American species were widespread over the Eurasiatic area. While the saber-toothed cats became extinct the world over, the group to which the great lionlike cat belongs still persists in the Old World and in South and Central America as well as southern North America.

In contrast to the sabertooth and American lion, the puma and lynx occurring at Rancho La Brea are closely allied to types still living in western North America. The living species of puma (*Felis concolor* Linnaeus) is included in the collection, and the extinct *Felis bituminosa* Merriam and Stock is now generally considered to be indistinguishable from the living puma. *Felis daggetti* Merriam is structurally similar to but slightly larger than the western mountain lion. The lynx, or bobcat, from Rancho La Brea, originally assigned to an extinct subspecies (*Lynx rufa fischeri* Merriam) on the basis of characters of its cheek teeth, is no longer considered distinct from the living lynx.

Jefferson (1983) described a partial palate, vertebrae, ribs, and limb bones that are referable to the large extinct subspecies of jaguar *Panthera onca augusta* (Simpson, 1941). These represent at least five individuals that were larger in size than most extant jaguar subspecies and range in age between 11,600 and 28,000 years. A right femur from Rancho La Brea that Merriam and Stock (1932) tentatively referred to *Felis daggetti* is probably better regarded as a specimen of *P. onca*, although one other specimen of that extinct species has been recovered from the asphalt deposits. The rarity of jaguar remains (20 specimens) at Rancho La Brea may reflect the relative abundance of the American lion. The two species appear mutually exclusive, the most abundant record of late Pleistocene jaguar being from areas such as peninsular Florida, Texas, and Tennessee, where lion is scarce or absent (Kurtén and Anderson, 1980).

RODENTIA (Mice, Rats, Squirrels)

Although remains of rodents are only sparsely represented in the Page Museum collection, these forms are known by a number of skulls and many parts of skeletons in the collection of the University of California, Berkeley. Many of the earlier excavations concentrated principally on the larger and more spectacular specimens. As a result, our knowledge of the Rancho La Brea rodent fossils was until recently largely restricted to specimens that had been fortuitously preserved in matrix adhering to the larger bones or that within the braincases of the larger fossil skulls. Many such micromammals were encountered during W. D. Pierce's investigation of fossil insects preserved in these endocranial cavities. The initial contributions to our knowledge of these forms were made by Kellogg (1912), Dice (1925), and by Wilson (1933). A considerable quantity of micromammal-bearing matrix was later recovered from the Pit 91 excavation but most has yet to be sorted under the microscope.

The surface activities of rodents during the Pleistocene were probably instrumental in bringing about a record of the group at Rancho La Brea. The relatively minor outpours of asphalt which occur today at this locality do on occasion catch these small mammals (see Figure 4). Furthermore, the occurrence of particular kinds of rodents doubtless accounts also for the presence of certain species of birds and small carnivorous mammals known to feed on living representatives of these forms.

The entire rodent assemblage, including not less than ten species, bears a close similarity to that living in the Los Angeles region at the present time. The southwestern pocket gopher (*Thomomys bottae occipitalis* Dice) is by far the most abundant rodent in the fossil collections from Rancho La Brea. The California pocket mouse (*Perognathus californicus* (C. H. Merriam)) and the Pacific kangaroo rat (*Dipodomys agilis* Gambel) are relatively abundant; the extinct imperfect mouse (*Peromyscus imperfectus* Dice), Californian voles (*Microtus californicus* (Peale) and *M. c. sanctidiegi* L. Kellogg), and California ground squirrel (*Spermophilus beecheyi* (Say)) are less common; and the southern grasshopper mouse (*Onychomys torridus ramona* Rhoades), western harvest mouse (*Reithrodontomys megalotis longicaudus* Baird), and dusky-footed wood rat (*Neotoma fuscipes*) are rare. The Rancho La Brea biota is also known to include Merriam's chipmunk (*Tamias* cf. *T. merriami*) (Whistler, 1989).

It is interesting to note that many of the rodents are referred to living species and even to existing subspecies. Although profound changes have occurred in the mammalian life of the Los Angeles area since the time of accumulation of the asphalt deposits, as indicated by the disappearance of many of the larger forms, the rodents apparently have remained remarkably stable, not only with reference to the constituent members of the group as a whole but also with regard to the structural characters of particular species.

LAGOMORPHA (Rabbits, Hares)

According to Dice (1925) at least three distinct types of lagomorphs are known from Rancho La Brea. These have been identified as the black-tailed jackrabbit (*Lepus californicus* Gray), the brush rabbit (*Sylvilagus bachmani* (Waterhouse)), and the desert cottontail (*Sylvilagus audubonii* Baird). Both the jackrabbit and the brush rabbit are represented by only a few bones, but the cottontail is relatively abundant in the asphalt. Today, the jackrabbit is widely distributed in prairie and desert habitats, the desert cottontail inhabits open plains, foothills, and valleys with grass sagebrush and juniper, whereas the brush rabbit occurs in dense chaparral (Kurtén and Anderson, 1980).

PERISSODACTYLA

EQUIDAE (Horses)

The presence of herds of horses in the vicinity of the asphalt deposits during the period of accumulation is clearly testified to by the numerous remains of these mammals found

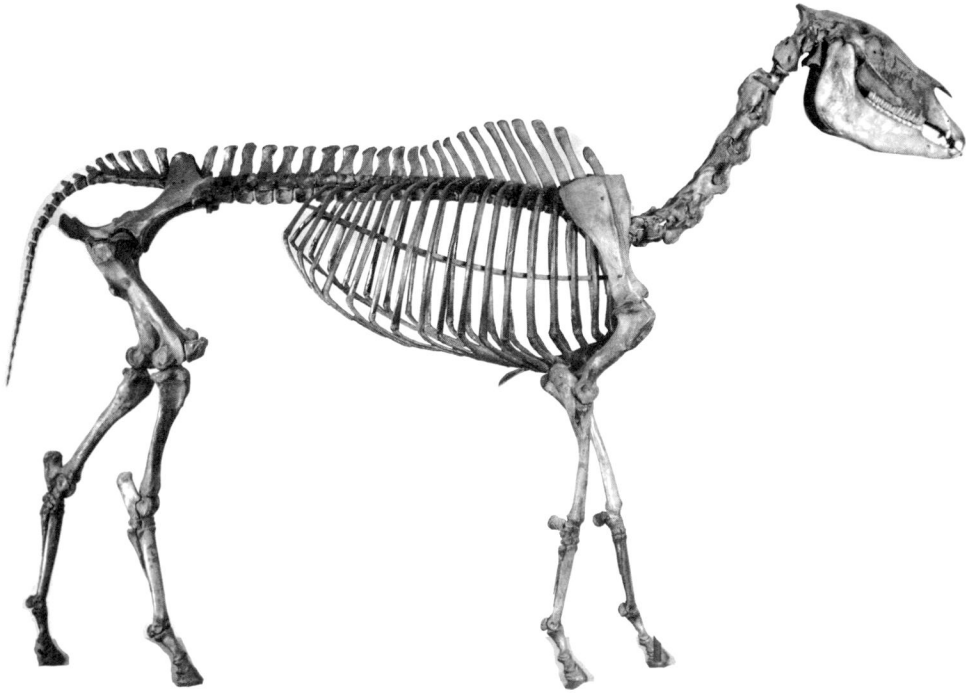

Figure 22. Composite skeleton of western horse (*Equus occidentalis* Leidy). Page Museum collection; Rancho La Brea Pleistocene.

at Rancho La Brea. Of the many individuals recorded in the collections, most belong to a single species that was tentatively identified as the extinct western horse (*Equus occidentalis* Leidy). In stage of evolution and in general body structure this extinct species resembles the modern horse, although differing from it in a number of specific details. Standing on the average about 14.5 hands (1.47 m, or 4 ft 10 in) at the withers, this animal was of the height of a modern Arabian horse. It was, however, of considerably heavier build.

Horses were among the more common types of hoofed mammals on the North American continent during Pleistocene time and several distinct species have been described from fossil remains. The abundance and widespread distribution of horses in North America make the apparent disappearance of the group in this region prior to the arrival of European explorers an added and an unusual feature of their long and eventful career.

The common large horse from Rancho La Brea (Figure 22) is now known to be represented by teeth of at least 220 individuals, most of which died during their first year of life (Scott, 1991a). Investigation of equid postcranial remains from radiometrically dated sites by Gust (1991) provided a minimum number of 120 individuals of which 48 were immature and 46 subadult. Males outnumbered females by nearly two to one although the relative frequencies of sex and maturity vary from sample to sample.

In this fossil species, the skull is somewhat domed in the region of the forehead and the sutures separating the individual bones of the skull in this area give a slightly different pattern from that seen in living species. In both of these characters the Rancho La Brea horse exhibits some resemblance to the asses. Another noticeable difference between *Equus occidentalis* and *Equus caballus,* as the modern domestic species is called, occurs in the front end of the lower jaw between the cheek teeth and the cropping teeth, being deeper in the fossil form. The dentition of the Rancho La Brea animal is not fundamentally different from that in modern species, but individual grinding teeth may have a simpler enamel pattern. Indeed, among the species of fossil horses described from the Pleistocene of North America, *Equus occidentalis* has one of the least complicated patterns developed by the enamel on the wearing surface of the tooth crown (Stock, 1930; Kurtén and Anderson, 1980).

The Rancho La Brea horses, like their living relatives, were one-toed animals. Limb and body are supported wholly by the enlarged third toe, while slender splintlike bones represent the elements which during an earlier history of the horse group were more fully developed and carried the second and fourth toes. The hoofs in the Rancho La Brea species are distinctly smaller and more slender than in the larger types of existing horses. In this respect again, a greater resemblance is seen to exist with the asses and zebras.

Despite the distinctive nature of its domed forehead and other cranial features, the large Rancho La Brea horse was originally referred to less complete remains of *Equus occidentalis* from Tuolumne County; the wisdom of this attribution has been extensively debated during the past two decades. The Rancho La Brea species is also known by many specimens that have been recovered from the asphalt deposits of McKittrick.

Most extant equids feed preferentially on grasses, but plant fragments retrieved from a juvenile large *Equus* tooth from Rancho La Brea were more than 50 percent dicotyledonous tissue (Akersten et al., 1988). Amino acid determinations from three Rancho La Brea *Equus* bones have yielded $\delta^{13}C$ ratios that suggested a diet of mainly C_3 plants (Marcus and Berger, 1984). The diet of a herd of feral horses (*Equus caballus*) from New Mexico was observed to vary seasonally from 37 to 63 percent dicotyledons (Hansen, 1976), whereas that of donkeys (*Equus asinus*) from the arid southwestern United States have been reported to include large quantities of non-monocotyledonous plants (Ginnett and Douglas, 1982; Seegmiller and Ohmart, 1981). Grasses were not abundant in the Rancho La Brea flora (Warter, 1976) and the equids would necessarily be opportunistic feeders under such circumstances.

A second, considerably smaller, species of horse is represented by a few isolated specimens referred to *Equus conversidens* (Scott, 1990). The smallest species of the American stout-limbed Pleistocene horses, *E. conversidens* was the commonest small horse in much of North America during the later Pleistocene (Kurtén and Anderson, 1980). Though evidently rare at Rancho La Brea, *E. conversidens* is relatively abundant at sites of comparable age in the Mojave Desert (Jefferson, 1988).

TAPIRIDAE (Tapirs)

The presence at Rancho La Brea of this interesting group is documented by two phalanges in the collection of the University of California, a single foot bone (ectocuneiform) in the Page Museum collections, and part of a lower jaw from a construction site adjacent to Hancock Park. The rarity of tapirs in the asphalt record confirms that the environment in and about the asphalt traps was not particularly favorable to this family.

Two late Pliocene and Pleistocene tapir species are now recognized from the western United States. The larger, *Tapirus merriami* Frick, is better represented at the upper Blancan and Irvingtonian localities of California, but Rancholabrean age specimens have been recovered from the La Habra Formation and Palos Verdes Sand in Orange County. The smaller *T. californicus* Merriam is more common in Rancholabrean localities of southern California. A right mandibular ramus recovered in 1985 during construction activity adjacent to Hancock Park undoubtedly belongs to *T. californicus*. The two phalanges from the University of California, Berkeley, collections and the ectocuneiform from Pit 77 in the Page Museum collections were tentatively referred to *T. californicus* on the basis of their small size (Jefferson, 1986a, 1989b).

ARTIODACTYLA

TAYASSUIDAE (Peccaries)

Peccaries are hoglike creatures representing the New World division of the pig-peccary group. Existing members of the family range from the region of northern Texas to Patagonia. The infrequent record of peccaries in the Pleistocene mammalian assemblage of California leaves much to be desired in our present knowledge of the extinct species. The most complete material thus far known from California has been collected at Rancho La Brea and constitutes a fragmentary skull and several limb elements. These specimens represent the genus *Platygonus*, which ranged widely over the United States and Mexico in Pleistocene time. Species closely related to the Rancho La Brea specimens have been described, for example, from fissure deposits in the lead region of Illinois, alluvial accumulations in northern Kansas, and from a number of additional localities in the East and Midwest. In California, the genus *Platygonus* has been recorded from the asphalt deposits of McKittrick in Kern County and doubtfully from the Pleistocene of Potter Creek Cave, Shasta County.

The characters of the skull and teeth of peccaries are in general like those in swine, although differing from them in a number of important details. The skull specimen from Rancho La Brea exhibits the typical features found in *Platygonus*. Two incisor teeth are present on each side of the upper jaw, followed behind by a short, triangular-shaped canine tooth. The cheek teeth have relatively low crowns on which a series of tubercles or cusps are developed. In the molars these cusps pair off to form two transverse crests which are, however, not so high as in modern peccaries.

Peccaries were represented during the Pleistocene in North America by two genera— *Platygonus* and *Mylohyus* (Kurtén and Anderson, 1980). *Mylohyus* was restricted mainly to the eastern part of the country but *Platygonus* was more widely distributed. Where the two overlap, as in Texas, *Mylohyus* is more common in assemblages of glacial age but *Platygonus* is more common in deposits known to represent interglacials or interstadials (Slaughter, 1966). However, *Platygonus compressus* LeConte, the species probably represented by the Rancho La Brea material, was a gregarious form with wide-ranging environmental tolerance (Kurtén and Anderson, 1980). According to Slaughter (1961, 1966) and Lundelius (1960), and in contrast to Stock's (1949) interpretation, *Platygonus* probably inhabited a variety of habitats including grassland, whereas *Mylohyus* preferred more wooded areas.

CAMELIDAE (Camels, Llamas)

Foreign as the camels appear to the North American mammalian life of today, this group of animals was well represented over the northern continental area of the New World during the Age of Mammals. Only late in geologic time have they disappeared entirely from this region, the group as a whole being now represented by the bactrian camel of Asia, the dromedary of northern Africa, and the llamas of South America.

Several distinct kinds of camels are known from the Pleistocene of North America and some of these were broadly distributed, ranging northward beyond the Arctic Circle. Remains of these creatures have been found at a number of localities in the United States from Washington to Florida.

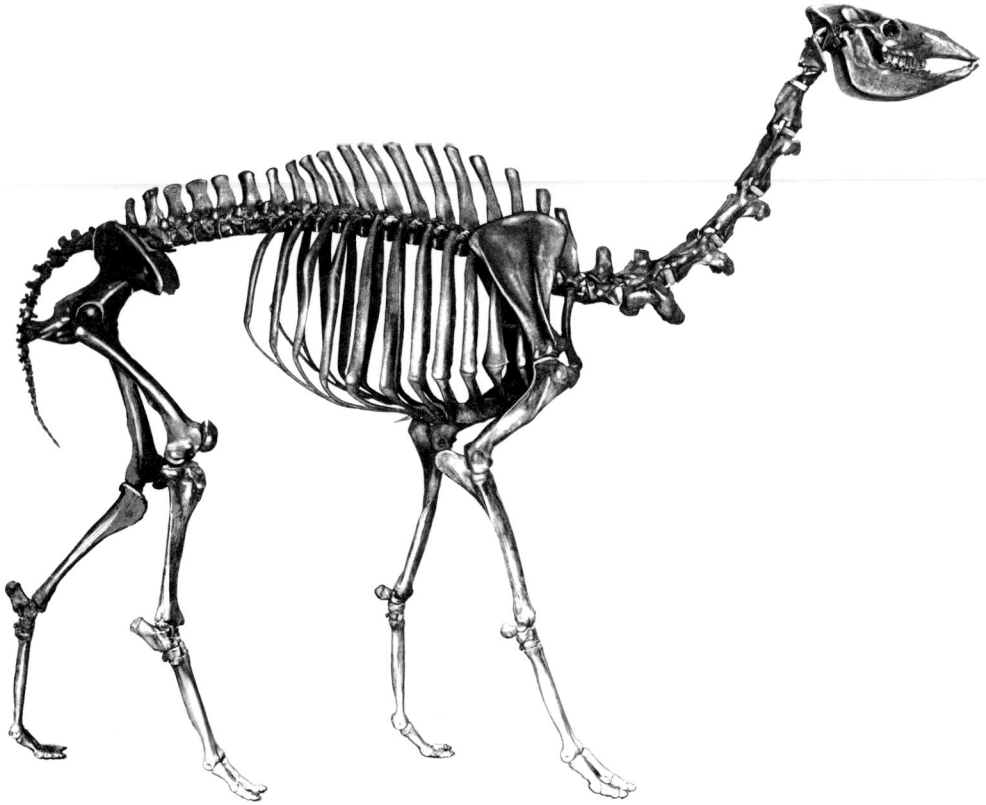

Figure 23. Composite skeleton of large camel (*Camelops hesternus* (Leidy)). Page Museum collection; Rancho La Brea Pleistocene.

The larger camelids of Rancho La Brea belong to the species *Camelops hesternus* (Leidy), whose presence in the vicinity of the asphalt seeps must indeed have been striking. The fossil materials furnish practically all the important structures of the skull and skeleton, the species being better known than any other camel type described from the Pleistocene. Two fine skeletons have been prepared and mounted from specimens collected by the Natural History Museum of Los Angeles County (see Figure 23). One is on display at the Page Museum, the other at the Natural History Museum in Exposition Park.

The mounted specimens of *Camelops hesternus* show the unusual characteristics of these fossil forms. The skeleton has a height of more than 2.13 m (7 ft) as measured from the highest point of the back, or more than 2.4 m (8 ft) measured from the top of the skull. While the size of the body was somewhat like that in the bactrian camel, the fossil species has much longer legs. Although the skull of *Camelops hesternus* agrees more in size with that of the bactrian camel than with that of the South American llama, it resembles more closely the latter in certain structural characters. However, there remain other features in which the Pleistocene species differs from the camelids now living in both the Old and the New World.

The species *Camelops hesternus* was first described in 1873 from beds in Livermore Valley, Alameda County. Skull remains of a camel identical with the Rancho La Brea form have been described by Romer (1925) from a cave near Fillmore, Utah. This specimen is remarkable because of its comparatively fresh state of preservation, for some of the dried muscle tissue is still attached to a part of the skull. Discovery of this interesting material, as well as of jaw and skeletal specimens of another extinct camel similarly preserved in Gypsum Cave, southern Nevada, has led to the belief held by some investigators that native camels continued to exist in North America during late geological time and perhaps into the Recent. A ^{14}C date of about 9,000 years, associated with *Camelops* material from Daggett, California (Reynolds and Reynolds, 1985), appears to support this assumption.

The definitive work on *Camelops* was furnished by Webb (1965). Although superficially closer to extant camels than extant llamas in overall size and shape, *Camelops* belonged to a different tribe of camelines that was in fact more closely related to the llamas. *Camelops* differed from extant camels and their close fossil relatives by its longer, relatively narrower, and proportionately deeper head in which the face flexed downwards more steeply than in the dromedary. The muzzle in particular was longer than in modern camels or llamas and the pendulous split upper lip was much heavier and better muscled. The neck vertebrae were proportionately longer than in living camels. The limbs were some 20 percent longer, the upper part of the limb being proportionately more elongate. The toes of *Camelops* were neither as flattened nor as symmetrical as those of living camels and more closely approached the condition seen in llama feet. Webb (1965) noted that in modern camels there is a correlation between the position of the hump and the relative height of the neural spines. On this basis, *Camelops* would have had a single hump running most of the length of the rib cage; the hump would have looked rather like that of the dromedary except for being located slightly farther forward.

Camelops and other fossil camelids from North America have often been interpreted as grass eaters on the assumption that their high crowned teeth were adaptations for a grazing diet. The extant dromedary (*C. dromedarius*) also possesses hypsodont cheek teeth but is an opportunistic feeder and its diverse diet includes a preponderance of dicotyledonous plants. Plant fragments extracted from Rancho La Brea *Camelops* teeth were less than 11 percent monocotyledonous plant tissue, suggesting that this extinct genus was also an opportunistic feeder (Akersten et al., 1988).

The presence of a second camelid species in the Rancho La Brea fauna—*Hemiauchenia macrocephala* Cope (the long-headed, stilt-legged llama)—is based on a few isolated limb bones (Miller, 1968) and vertebrae that are distinctly smaller than those of the more abundant *Camelops hesternus*. Unpublished field notes of the 1913–1915 excavations by L. E. Wyman indicate that a skull of a "new llama-like camel" was discovered in 1914 but unfortunately this specimen was not recoverable (W. E. Miller, 1968). Species of *Hemiauchenia* were larger than extant llamas, from which they also differed by having proportionately longer limbs and much longer metapodials. The stilt-legged llamas were cursorially adapted for life in the open grasslands of North and South America, the shorter legs of their extant descendants being adaptations for the more rugged terrain of the South American highlands (Webb, 1974). Although rare at Rancho La Brea, *Hemiauchenia* was relatively common at McKittrick and Pleistocene localities in the Mojave Desert (G. T. Jefferson, pers. comm.). In contrast to *Hemiauchenia*, *Camelops* was restricted in distribution to North America.

CERVIDAE (Deer)

Although the deer are now the most abundant of the larger game animals of California, and have been recorded from a number of Pleistocene localities, they occur only rarely in the Rancho La Brea Pleistocene assemblage. Judging from the rather scanty material available, the Rancho La Brea specimens are related to the California mule deer (*Odocoileus*

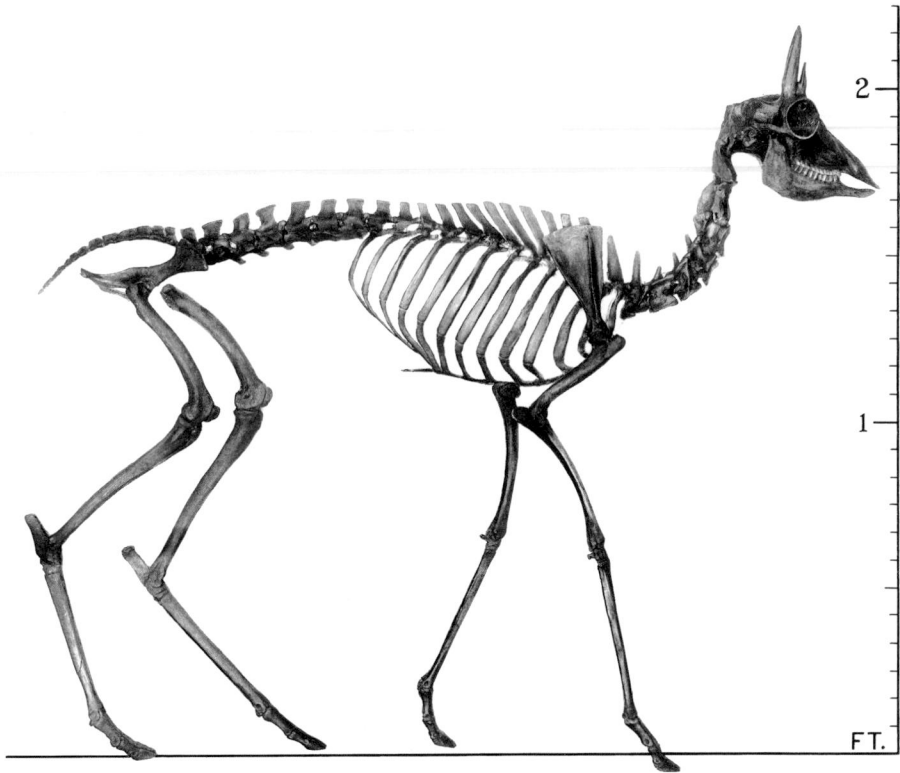

Figure 24. Composite skeleton of small antelope (*Capromeryx minor* (Taylor)). Page Museum collection; Rancho La Brea Pleistocene. After Furlong.

hemionus (Rafinesque)), although establishment of definite relationships of these forms must await the recovery of further and more diagnostic fossil specimens. The rarity of deer in the asphaltic deposits may indicate an infrequent occurrence of the group during this stage of the Pleistocene, or the lack of environmental conditions favorable to these animals. The American elk, or wapiti, (*Cervus elephas* Linnaeus) is known in artifactual context by a few specimens from the Holocene portion of the sequence.

ANTILOCAPRIDAE (Pronghorns)

This distinctly American family is represented today by a single species, the pronghorn (*Antilocapra americana* (Ord)), now largely restricted in its range to certain areas in the western part of the United States and northern Mexico. Prior to the arrival of European settlers, however, its distribution extended over a much larger area. Occurring also in California during the Recent epoch, its range in this region has likewise become much more restricted, and only a few herds are left. Fossil remains of species identical with or closely related to the pronghorn have been found in Pleistocene deposits at several localities in California. Large antilocaprid specimens from Rancho La Brea and McKittrick cannot be distinguished from *Antilocapra americana* although a different species (*A. pacifica* Richards and McCrossin) has been recognized from a Rancholabrean aged locality in northern California (Richards and McCrossin, 1991).

Much more abundant in the asphaltic sediments, however, is a diminutive pronghorn (*Capromeryx minor* (Taylor)) which clearly belongs to the antilocaprid family, although exhibiting some interesting differences from the American pronghorn of today. This animal (Figure 24) was less than 60 cm (24 in) tall at the shoulders or approximately 68 cm (27 in) as measured to the top of the head. It possessed tall-crowned grinding teeth, suggestive of dietary habits similar to those of the living pronghorn. While no complete skull has been found at Rancho La Brea, sufficient material is available to permit determination of the character of the bony horn core. This interesting structure, growing on each side from the top of the skull above the rim of the eye socket, consists of two distinct prongs arising from a common base. The hinder of the two prongs is the longer and has a round cross-section, while the front prong is decidedly shorter and has a triangular cross section. Although the swordlike horn core of the modern pronghorn antelope shows no separation into two parts, the contour and shape of this structure are rather strongly indicative of a kinship between the modern type and *Capromeryx*. Moreover, in the fossil form the cleft separating the two prongs suggests the presence of a forked sheath covering the horn core. The existing pronghorn is the only known case among living mammals in which a simple horn core carries a forked sheath.

Taylor (1911) was uncertain of the generic attribution of the diminutive pronghorn he described from Rancho La Brea, and for which Furlong (1946) proposed the descriptive name *Breameryx minor*. Most subsequent workers have, however, included the diminutive Rancho La Brea pronghorn with three other small antilocaprids in the genus *Capromeryx* (Kurtén and Anderson, 1980). There are a number of resemblances in the skeletons of *Capromeryx* and *Antilocapra*. Like the latter, *Capromeryx* possessed long, light limbs, which distinguish it from most of the true antelopes of the Old World. The taller-crowned teeth serve to distinguish it from the light-limbed African gazelles.

The forked horn sheath of living male pronghorns is a relict of the bony bifurcation of ancestral forms such as *Antilocapra* (*Subantilocapra*) *garciae* from the late Pliocene of Florida (Webb, 1973). Although it is not closely related to *Antilocapra*, the horns of *Capromeryx*, with their short anterior and longer posterior forks, are thus more typical of the family Antilocapridae than are the horns of the living representatives.

The presence of delicate creatures like *Capromeryx* in a region in which so many formidable carnivorous animals also existed seems somewhat incongruous. Perhaps these antelopes normally took shelter in the copses of trees and shrubs in the vicinity of the asphalt seeps, coming into the open only at certain times of the day or night to feed. Or perhaps, like the small Thomson's gazelle of the East African savannas, their association in large herds on open plains assured survival in the presence of powerful predators.

Specimens assigned to the genus *Capromeryx* have been recognized, in spite of the rather fragile nature of the remains, in Pleistocene deposits of New Mexico and of the Valley of Mexico. Additional records of the presence of this group of animals in the California region are known from Pleistocene beach accumulations at San Pedro, from the asphalt deposits at McKittrick, Kern County, from localities in the Mojave Desert, and doubtfully from the Bautista Creek badlands of Riverside County.

BOVIDAE (Bison, Shrub Oxen)

Judging from their record in the asphaltic sediments, the bison, or buffalo, were even more numerous than the horses in the vicinity of Rancho La Brea during the Pleistocene. The total number of these animals exceeds that of all other even-toed hoofed mammals (camels, peccaries, antelopes, and deer) found at this locality. It should be clearly borne in mind, however, that factors other than abundance of individuals may have been instrumental in bringing about the large representation of the bovid group in the asphalt. The extinct bison, with habits similar to those of the modern species, may easily have fallen victim to the deceptive appearances of the asphalt seeps. Two distinct species have been recognized from the asphalt deposits: the common ancient bison (*Bison antiquus* Leidy) and the larger but rarer long-horned bison (*Bison latifrons* (Harlan)). The similarity in most fundamental characters of the fossil forms to the living species makes it easy to visualize these creatures in the flesh, roaming over the plains and rolling country of the Los Angeles area during the Pleistocene.

Wyman (1926) recorded the presence of a large bison skull from Rancho La Brea that had horn cores measuring "a full six feet from tip to tip." Unfortunately this specimen disintegrated while being taken out of the ground. Very large bison postcranial bones were recovered from Pits 3, 4, and 9 and were subsequently attributed to *Bison latifrons* (W. E. Miller, 1968; Miller and Brotherson, 1979). The systematics of North American bison species is currently not well understood, although the migration of the genus from Asia to North America characterizes the beginning of the Rancholabrean Land Mammal Age. It seems entirely possible that *Bison latifrons* constituted the ancestral form from which the smaller *B. antiquus* evolved. Indeed Lundelius et al. (1987) suggested that *Bison antiquus* may represent a smaller-horned end member of the *B. latifrons–B. alleni* lineage. The largest known specimen of *Bison latifrons* is a complete skull from an ancient lake bed in Shasta County. In this specimen the spread of the horns measured more than 1.98 m (6.5 ft) from tip to tip.

Remains of *Bison antiquus* are present in Pleistocene localities throughout California. The species was recorded many years ago from deposits in Livermore Valley, Alameda County, California, and is also represented in the asphalt deposits of McKittrick. During its existence on the North American continent *Bison antiquus* was widely distributed, for the species was first described from the Pleistocene of Kentucky.

Bison antiquus was larger than the modern North American buffalo (*Bison bison* Linnaeus), the skeleton as mounted (Figure 25) having a height of more than 2.2 m (7 ft), measured from the highest point of the back. Some individuals were probably distinctly taller. That these fossil bison, like their living relatives, possessed a hump is indicated by the great development of the spines of the vertebrae in the forward region of the back. Vertebrae are available in the collection with spines measuring from 66 to 68 cm (26 to 27 in) in length.

Structurally, *Bison antiquus* resembles *Bison bison*, but differs from it in details other than size. In the skull, for example, the horn cores project outward at right angles to the median fore and aft axis of the head, whereas in the living species they are directed somewhat backward as well as outward. The largest skull in the museum collection has a span of 81 cm (32 in) as measured between the tips of the horn cores. It is not improbable that some individuals possessed a width across the skull of more than a yard as measured from the outer sides of the horny sheaths which covered the horn cores. Skulls of males and females are apparently to be distinguished by the size and shape of the individual horn cores.

Bison antiquus is represented by at least 300 individuals, females outnumbering males by a ratio of seven to two. Most of the bison specimens, like those of the horses, represent juvenile animals (288 individuals: Jefferson and Goldin, 1989). However, the juvenile dentitions represent specific age clusters (individuals of 2–4 months, 14–16 months, 26–28 months, etc.) rather than a continuous growth series and would appear to signify the seasonal presence of female bison, their calves and other juvenile bison in the area two months after the peak calving season (Jefferson and Goldin, 1989). A similar distribution could be expected from seasonal variation in the entrapment mechanism, but Jefferson

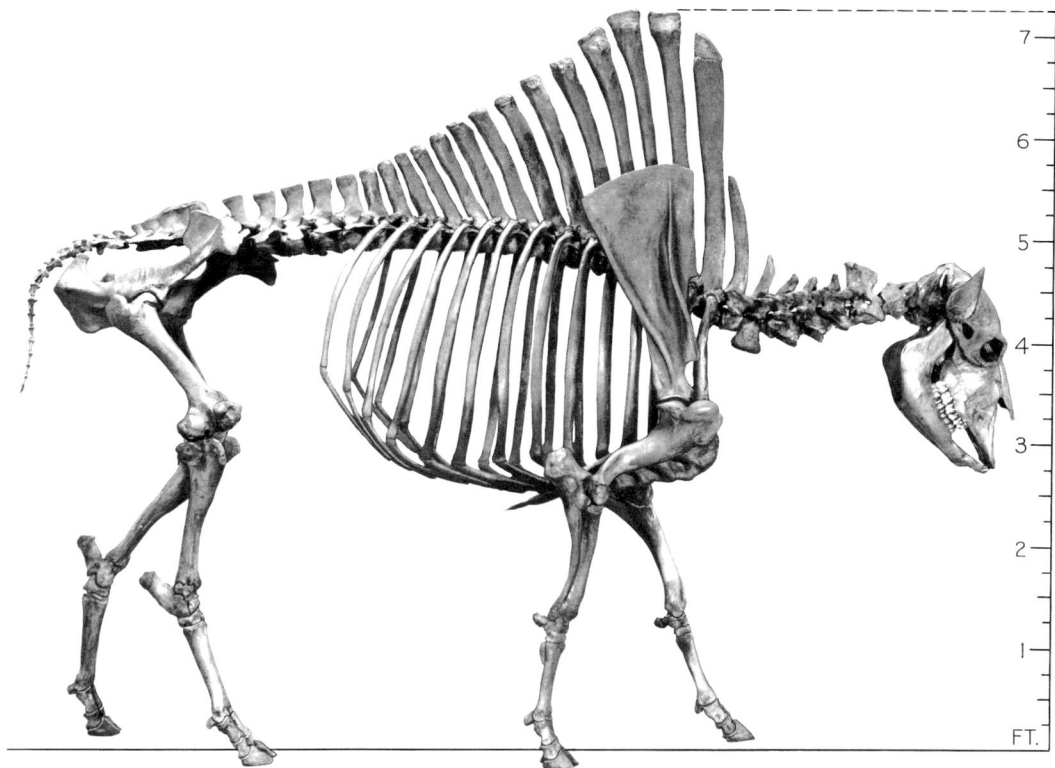

Figure 25. Composite skeleton of ancient bison (*Bison antiquus* Leidy). Page Museum collection; Rancho La Brea Pleistocene.

and Goldin (1989) interpret calving to have taken place at some time during late spring (April and May) and the peak fluidity of the asphalt does not occur until the late summer (August and September).

The extant *Bison bison* feeds almost exclusively on grasses and sedges (Van Vuren, 1984) but plant fragments extracted from depressions in the teeth of *B. antiquus* from Rancho La Brea contained only 13.4 percent of monocotyledonous plants, contrasting with 75 percent of monocotyledonous plant material from dental boluses in the teeth of living bison (Akersten et al., 1988). It is possible that, like the European wisent (*Bison bonasus*), *Bison antiquus* fed mainly on dicotyledonous plant materials. Many of the plants represented in the boluses are xeric species and perhaps were not eaten locally. However, grasses are rare in the diverse flora of Rancho La Brea (Warter, 1976) and it is possible that the composition of plant remains preserved in the *B. antiquus* teeth reflects the preponderance of available vegetation.

Although not yet recovered from within Hancock Park itself, remains of the shrub ox (*Euceratherium collinum* Furlong and Sinclair) were encountered in asphaltic deposits during construction activities immediately adjacent to the park. Distantly related to the musk ox (*Ovibos*) and woodland musk ox (*Symbos*), *Euceratherium* was restricted to the western and southwestern United States (Lundelius et al., 1987). Known from only one species, *Euceratherium* was a large specialized ovibovine that probably inhabited lower hills (Kurtén and Anderson, 1980).

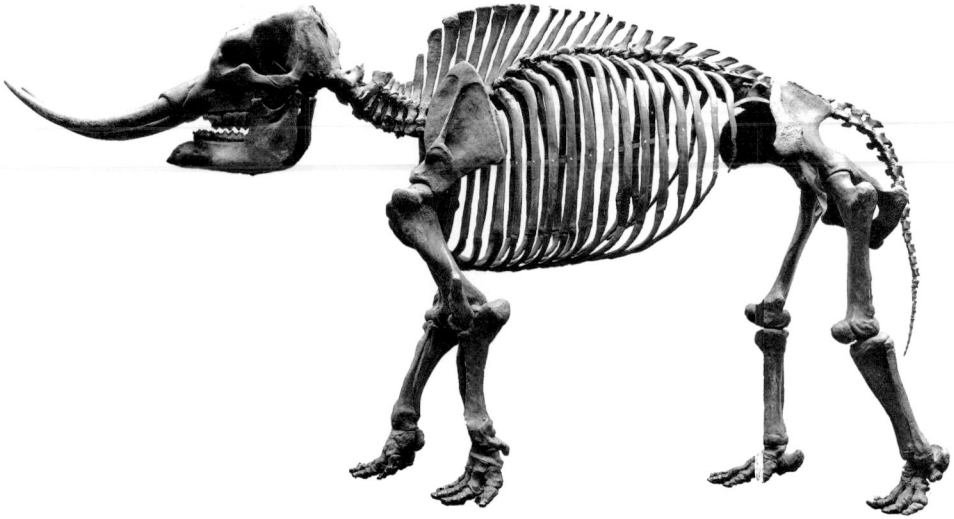

Figure 26. Composite skeleton of American mastodont *(Mastodon americanum* (Kerr)). Page Museum collection; Rancho La Brea Pleistocene.

PROBOSCIDEA

The American mastodont is known to have ranged during Pleistocene time from Alaska to Florida and from New England to southern California. The Columbian mammoths were similarly widespread in their distribution. Remains of Pleistocene mastodonts and mammoths have been found at a number of localities in California, but recognition of more than one species of mammoth during the last glacial episode is probably unfounded.

Mammoth and mastodont remains are by no means common in the Rancho La Brea collections. Moreover, in contrast to the other large herbivores, many of the proboscidean limb bones are imperfectly preserved. This is particularly true for the mammoth remains collected by the Natural History Museum from the excavation at Pit 9. Here the presence of clay and sand saturated with water had hastened the disintegration of the skulls and the long bones of the limbs. In contrast, remains of mastodont encountered at other Rancho La Brea localities are usually in a better state of preservation.

MASTODONTIDAE (Mastodonts)

The contemporaneous occurrence of the mastodont and mammoth is clearly indicated in the brea deposits. The former, determined as the American mastodont *(Mammut americanum* (Kerr)), was of smaller size than the mammoth and possessed a number of primitive features, but exhibited, nevertheless, many of the outward characteristics of the proboscideans. A mounted specimen of this species from the asphalt is shown in Figure 26. This skeleton measures 1.9 meters (6 feet 3 inches) in height to the top of the shoulder blade. The tusks of the mastodont, made of dentine or ivory, are of smaller size than those of the Pleistocene mammoths and their curvature is not so marked as in the latter. A greater number of grinding teeth are present in the jaws of mastodonts and each individual tooth differs noticeably in its structure from that of the mammoth. The name "mastodont" means "nipple tooth," which aptly describes the shape of the low large cusps arranged in pairs on each tooth. Mastodont teeth are comparatively low-crowned, each crown forming a series of crests and V-shaped valleys which extend transversely. There is no cement on the wearing surface of the grinders in the American mastodont.

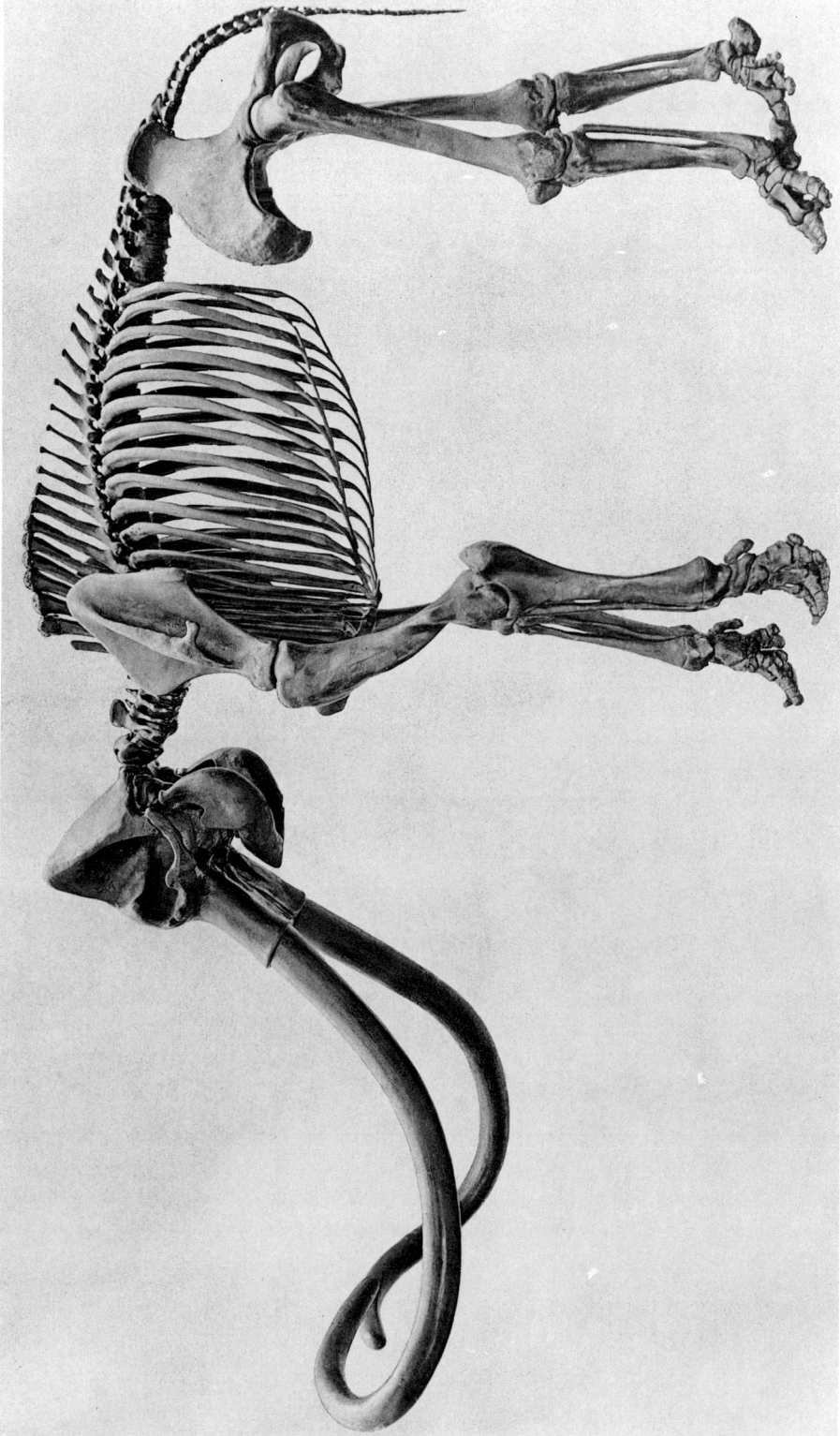

Figure 27. Composite skeleton of Columbian mammoth (*Mammuthus columbi* (Falconer)). Page Museum collection; Rancho La Brea Pleistocene.

The preferred environment of mastodonts is thought to include open spruce woodland and spruce forests, and analysis of undigested plant remains found in the rib cages of some specimens has revealed twigs and cones of conifers, leaves, coarse grasses, swamp plants, and mosses (Kurtén and Anderson, 1980; Lepper et al., 1991). Mastodonts were not, however, restricted to one habitat and those present at Rancho La Brea would have exploited any available browse. At least 14 individuals are known from the Rancho La Brea deposits, 5 of which were associated with mammoth remains from Pit 9. Mastodonts have not been recorded from the more arid late Pleistocene localities of southern California (Jefferson, 1988a).

ELEPHANTIDAE (Mammoths)

The mammoths of the asphalt were distinctly larger than the mastodont, and exceeded in size their living elephant relatives. Some of the animals had a height of more than 3.3 m (13 ft) as measured at the shoulders. The skeleton of *Mammuthus columbi* (Falconer) originally mounted in the Hancock Hall of the Natural History Museum, Figure 27, measures in height 3.26 m (10 ft 8.5 in) to the top of the shoulder blade, and 3.53 m (11 ft 7 in) to the top of the skull. The tusks were huge structures and were present only in the upper jaw. The grinding battery in adult animals consists of a single large tooth situated on each side of the upper and lower jaw. As in modern elephants, the crown of each tooth is made up of a series of compressed enamel plates enclosing dentine and virtually embedded in a heavy deposit of cement. The long-crowned teeth exhibit great wearing qualities and undoubtedly contributed to the longevity of these animals. The large skull furnished an expanse of external surface for attachment of muscles and tendons necessary in the manipulation of the trunk and in the support of the tusks. The heavy body was supported by the enormous pillarlike limbs.

Stock (1949) believed that both the imperial mammoth (*Mammuthus imperator* (Leidy)) and the Columbian mammoth (*Mammuthus columbi* (Falconer)) were present at Rancho La Brea but only one species (*Mammuthus columbi*) is represented in the Page Museum collections. Remains of at least 16 individuals were recovered from Pit 9, and 2 other specimens were collected from Pit 17 (Shelley Cox, pers. comm.). The restricted distribution of this species in the Rancho La Brea assemblages has yet to be satisfactorily explained. Perhaps mammoths were present in the Los Angeles Basin only during part of the interval represented by the total stratigraphic sequence. Radiometric dating of mammoth bone from Pit 9 has not been successful; fossil wood from that locality has yielded an estimate of nearly 40,000 years BP, although this date is considered unreliable. Alternatively, perhaps the circumstances that combined to permit the entrapment of such large and intelligent mammals occurred only rarely. Two factors are relevant to this possibility. First, elephantids were better adapted for grazing than their other proboscidean relatives, and mammoths were the most highly evolved elephantids in this respect. If mammoths were present in the basin only during the annual flush of grass accompanying the rainy season, their presence would precede the onset of maximum asphalt flow at the beginning of summer. Second, the field notes (and the waterlogged nature of recovered bone) suggest that Pit 9 was located at the site of a spring, which could have served as a focus for water-dependent large mammals. Surface asphalt adjoining or beneath the surface water would enhance the susceptibility of incautious or physically distressed mammoths.

XENARTHRA (Ground Sloths)

The ground sloths are among the most unusual kinds of herbivores occurring in the asphaltic deposits. Belonging to the edentates, an important but rather primitive group of mammals, these forms are known to have emigrated from the South American continent to North America. Forerunners and close relatives of some of the Rancho La Brea forms had already reached the northern continent during the Pliocene, the epoch immediately preceding the Pleistocene. Living representatives of the Xenarthra, which are found today in South and Central America, include the anteaters, tree sloths, and armadillos. Armadillos occur also in North America, but their geographic range is largely restricted to certain parts of Texas and Oklahoma.

Among existing edentates the nearest relative of the extinct ground sloth is the tree sloth. This, as its name implies, is a tree-dweller characterized by sluggish movements; it lives in the dense forests of Central and northern South America. In contrast to other arboreal mammals, the tree sloth has the peculiar habit of moving about with its body suspended from the branches of trees, using for support the large claws on its hands and feet. Of relatively small size, it feeds on the foliage of the trees, and when forced to the ground moves about with difficulty. The tree sloth possesses a shaggy coat of hair and in some forms this has a protective color, due to the presence of a green alga which lives normally in the flutings or grooves of the individual hair shafts.

The extinct edentates of the Western Hemisphere include not only the ground sloths, but forms characterized by a heavy bony armor which protected the head and encased the body and tail. The glyptodonts, as they are called, were like the armadillos in possessing this protective covering of bony dermal plates, or scutes. Glyptodonts are not known in the fossil record of California.

Some of the edentates, notably the anteaters, are toothless. When a dentition is present, as in some forms, the teeth lack the hard outer layer of enamel which is an important constituent of the tooth crown in most higher mammals. The ground sloths had teeth of this type, in which the principal substance forming the crown was dentine of varying hardness. As the teeth were worn down they continued to grow from within the sockets, very much as do the gnawing teeth of rodents.

MYLODONTIDAE (Grazing Ground Sloths)

Three distinct species of ground sloths are known from the Rancho La Brea deposits. Of these forms *Glossotherium harlani* (Owen) was the commonest type. This creature, represented in the collections of the Page Museum by very complete skull and skeletal materials from more than 60 individuals (Sherri Gust, pers. comm.), was much smaller than the mastodont, but, like that species, possessed considerable bulk and weight. Somewhat blunt-nosed and with lobate grinding teeth, *Glossotherium* was probably a grazing mammal frequenting the open stretches of flat or rolling country in the vicinity of the asphalt beds. Presumably its natural enemies were the sabertooth, the great lionlike cat, and the packs of dire wolves. The skeleton of this ground sloth is massively constructed, and the strongly built chest and powerful front limbs suggest great crushing strength (see Figure 28). These characters together with the stout claws of the hand were undoubtedly of great service to the animal in combating the attacks of predatory mammals.

An additional protection, particularly against attacks to the neck and shoulder region, was furnished by the nodules of bone, or dermal ossicles as they are called, that are embedded in the deeper layers of the skin. These bony elements have the same origin as the scutes in the armadillos and glyptodonts. While the skin or hide of *Glossotherium* is not preserved in the asphalt, the ownership of these nodules is indicated by the fact that they were found frequently in great abundance lying in the asphalt immediately adjacent to the skeletal remains of *Glossotherium*. Moreover, our knowledge of the structure of the skin in the mylodont ground sloths has been enriched by a discovery made many years

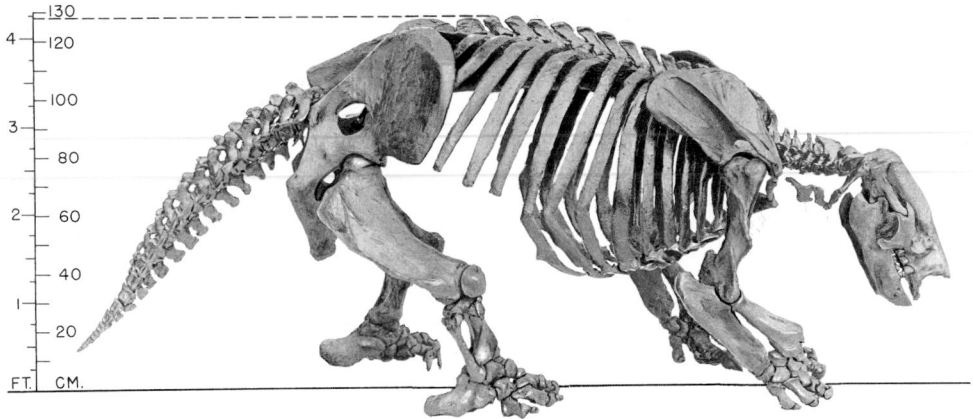

Figure 28. Composite skeleton of mylodont ground sloth *(Glossotherium harlani* (Owen)). Page Museum collection; Rancho La Brea Pleistocene. After Stock.

ago of remains of creatures closely related to the Rancho La Brea forms in a cave deposit in southern Patagonia. Here occurred not only the skull and skeletal remains, but patches of the hide, showing the presence of dermal ossicles in the deeper layers of the skin and the coarse shaggy hair on the outside. This discovery has had an important bearing on the interpretation of the characters of the Rancho La Brea mylodonts and on the restoration of these curious creatures.

The mylodont ground sloths enjoyed an extensive distribution over the American continent during the Pleistocene, their remains having been recorded (usually in association with those of plains-dwelling mammals) from the northern United States to Patagonia. In California, *Glossotherium* is known to have ranged from the Klamath River region southward to the Los Angeles Basin and skull or skeletal materials have been found at a dozen or more localities within the state. It is now well established that what were once interpreted as "human tracks" preserved in the silts and sandstones exposed in the prison yard of the Nevada State Penitentiary at Carson City, Nevada, represent the footprints of this kind of ground sloth (LeConte, 1882; Marsh, 1883; Stock, 1920b).

The cranial morphology and dietary adaptations of *Glossotherium harlani* from Rancho La Brea were investigated by Naples (1989). In comparison to extant tree sloths and to the browsing ground sloth, *Nothrotheriops shastensis* Sinclair, the cheek teeth of *Glossotherium* are proportionately larger, are more complex in shape, and retain a surrounding layer of cementum. The teeth of *Glossotherium* were less efficient for grazing than those of horses, camels, antilocaprids, or bison, but other factors such as very large body size but low metabolic rate, an unusually capacious gut (with a foregut fermentation site), and relatively slow passage of ingested food through the gut as in extant tree sloths, enabled this grazing ground sloth to compete successfully in the same habitat. Although the diet of *Glossotherium* probably contained a larger proportion of grasses than that of other contemporary ground sloths, this animal probably also ate a variety of foliage plants and might be better considered as a browser-grazer than a "pure" grazing animal (Naples, 1989).

A variety of this species, *Glossotherium harlani tenuiceps* (Stock), was also recorded from Rancho La Brea. *Glossotherium harlani tenuiceps* has a proportionately longer and more slender skull than *G. h. harlani* but these two subspecies are now believed to be female and male, respectively, of *Glossotherium harlani* (McDonald, 1991).

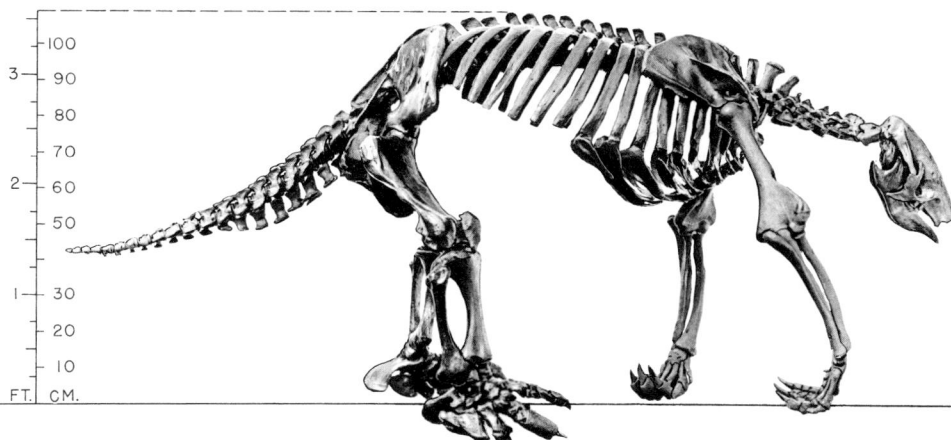

Figure 29. Composite skeleton of browsing ground sloth *(Nothrotheriops shastensis Sinclair)*. Page Museum collection; Rancho La Brea Pleistocene. After Stock.

MEGATHERIIDAE (Browsing Ground Sloths)

Next in abundance were the nothrothere ground sloths (*Nothrotheriops shastensis* Sinclair), which were smaller in size and of somewhat different appearance and habits than *Glossotherium*. The specimen shown in Figure 29 is unique not only because it is largely constructed of skeletal parts of a single individual, but likewise because the mounted skeleton represents the first of its kind to be attempted. At the time the materials were assembled for mounting, *Nothrotheriops* appeared to be so little known as a member of the North American company of Pleistocene mammals as to suggest the retention of its individual skeletal parts in a state which would permit their future study and comparison. As a result, a plaster replica was made of each element and these were displayed instead of the originals. However, subsequent explorations at new localities, as for example at Aden Crater, New Mexico, and in the cave of San Josecito, Nuevo León, Mexico, have brought to light excellently preserved remains of this ground sloth.

The nothrotheres are characterized by a fewer number of teeth than *Glossotherium* (eight above and six below) and the individual tooth was not lobate as in *Glossotherium* but rectangular in outline. Chisel-like edges on the teeth indicate a cutting or chopping rather than a grinding action in chewing food and suggest a browsing habit. The skull was somewhat tubular in front and the lower jaw has a spoutlike forward end.

Naples (1987) investigated the cranial anatomy and functional morphology of *Nothrotheriops shastensis*. Its relatively elongate face, long tongue, and flexible upper lip expedited the selective browsing of leafy materials. As in *Glossotherium*, a low metabolic rate and increased capacity for foregut fermentation would have compensated for the large body size and enabled *Nothrotheriops* to compete successfully with other selective mammalian browsers of the region. Although differing considerably in size and cranial characters, *Nothrotheriops* is evidently more closely related to the extant three-toed sloth (*Bradypus cuculliger cuculliger* Wagler) than to other living or extinct sloth genera (Naples, 1987).

Fundamentally, the *Nothrotheriops* skeleton is like that of *Glossotherium*, although differing in a number of details. Rudimentary bony nodules were apparently absent from the skin. Perhaps the most remarkable feature is presented in the hind foot, which has undergone profound modification. Whereas in the plantigrade mammals the sole of the foot rests on the ground, in *Nothrotheriops* the foot has been rotated from this position and the weight of the body rests on the outer side of the foot. As a result of this curious change, the outer toes particularly have suffered considerable reduction. A similar modification occurs also in the hind foot of *Glossotherium*, but is less striking than in the nothrotheres. Claws are present on both the feet and hands, as in *Glossotherium*.

It is difficult to conceive of these mammals moving with any degree of rapidity over the surface of the ground. Their ungainly appearance and peculiarities of skeletal structure suggest they were capable only of a slow and labored gait. Yet these characteristics were apparently not a handicap in their distribution, for the nothrotheres are found in Pleistocene strata of both North and South America. Browsing ground sloths were first described in California from cave accumulations in Shasta County. They have since been recognized at Hawver Cave in El Dorado County, California, and in Nevada, Arizona, Texas, New Mexico, and Mexico. The bones of *Nothrotheriops* from Gypsum Cave, Nevada are quite recent-looking with bits of dried integument and hair still associated with them. Whether these ground sloths lived in the region contemporaneously with humans is still an open question, but there is much evidence to indicate that they survived in the Southwest into early Recent time.

Studies of the plants in the dung of *Nothrotheriops* from the dry caves in Nevada and Arizona indicate that this animal fed on yucca and other desert plants and lived in a rather arid environment, like that found today in the Joshua tree forests of the Clark Mountains of Nevada. They do not appear to have been habitual residents of Rancho La Brea, and may have penetrated this region on occasion from adjacent areas.

MEGALONYCHIDAE (Flat-footed Ground Sloths)

A third kind of ground sloth (*Megalonyx jeffersoni californicus* Stock) is more sparingly represented in the asphaltic deposits than *Nothrotheriops*. This form is more closely related to the nothrotheres than to *Glossotherium*, but more nearly approaches the latter in size. *Megalonyx* possesses considerable historic interest, for large claws and other bones of this creature were first described in 1794 from a limestone cavern deposit in western Virginia by Thomas Jefferson. Since that time, *Megalonyx* has been recognized in Pleistocene beds at a number of localities in the United States, often in association with forest-dwelling mammals. Specimens of *M. jeffersoni* from the southern part of the United States are smaller than those from the north. In California this ground sloth is known also from Pleistocene cave accumulations in Shasta County and in the Sierran region, and from the late Pleistocene Palos Verdes formation of the San Pedro Hills. The Palos Verdes material was originally attributed to *Megalonyx milleri* Lyon but is no longer considered distinct from *M. jeffersoni*. No skull of this species is known from Rancho La Brea, but a lower jaw and a number of scattered skeletal elements have been recovered.

Megalonyx is quite similar in structure to the nothrotheres. The teeth of *Megalonyx* are similar in shape to those of the smaller form, but there is a caniniform tooth at the front end of each tooth row in the upper and lower jaws which is absent in *Nothrotheriops*. The skull of *Megalonyx*, as known from other localities, appears to terminate more bluntly in front than does that of the nothrotheres. Unlike the other Rancho La Brea ground sloths, *Megalonyx* had a plantigrade hind foot and its weight was borne on the sole, rather than the side, of its foot. The three central claws of the hind foot were well developed and touched the ground, whereas in *Nothrotheriops* and *Glossotherium* these claws were smaller and elevated above the ground surface. *Megalonyx* apparently also lacked bony nodules in its skin, but was probably covered by a heavy coat of coarse hair. The largest known species of *Megalonyx*, Jefferson's ground sloth dwelt primarily in woodland and forest, where it browsed on leaves and twigs (Kurtén and Anderson, 1980).

BIRDS

The rarity of birds in the geologic record is in large measure attributable to the fragile nature of their remains and to the special conditions of rapid entombment that are frequently necessary for their preservation. There appears to be no particular reason for assuming that birds of flight were less numerous during later geologic time than at present. On occasion quite complete remains are encountered as fossils in sedimentary deposits. The occurrence of fragmentary specimens in formations representing several epochs of the Age of Mammals gives clear evidence of the existence of many kinds of birds, some differing widely from living species and others indistinguishable from modern types. The geographic distribution of extinct species of birds, however, is not so well defined as that of extinct mammals.

The fossil birds from Rancho La Brea form a more varied and certainly no less interesting assemblage than the fossil mammals (see Figure 30). More than 140,000 specimens representing over 130 different species of birds are now known from the asphaltic deposits. The fossil bird assemblage includes many species that cannot be distinguished specifically from their living relatives. Some were formerly referred to living species, but after further study are now regarded as slightly, but significantly, different. Some are now considered to be the direct ancestors of closely related living species. Still others known principally from Rancho La Brea, but subsequently identified at other Pleistocene localities, are so greatly different from modern birds as to clearly represent extinct forms. So abundant are the skeletal remains in the asphalt deposits that they can be reassembled into composite mounted skeletons of various different species. This is one of the many unique features of the collection on display at the George C. Page Museum.

The conditions responsible for the remarkably complete record of the mammalian life of the region were no less favorable in bringing about a full representation of the birds. Once in the vicinity of the asphalt traps, a chance contact of wings or feet with the sticky material made the danger of miring in the asphalt extremely great. Even the more powerful birds of flight obviously suffered a decided disadvantage when their principal means of escape were rendered useless. For a bird thus caught, further struggle would merely cause other parts of its body to be entrapped.

The record of different kinds of birds in the Rancho La Brea fossil assemblage is considered to be related to their abundance in the vicinity. Undoubtedly the presence of

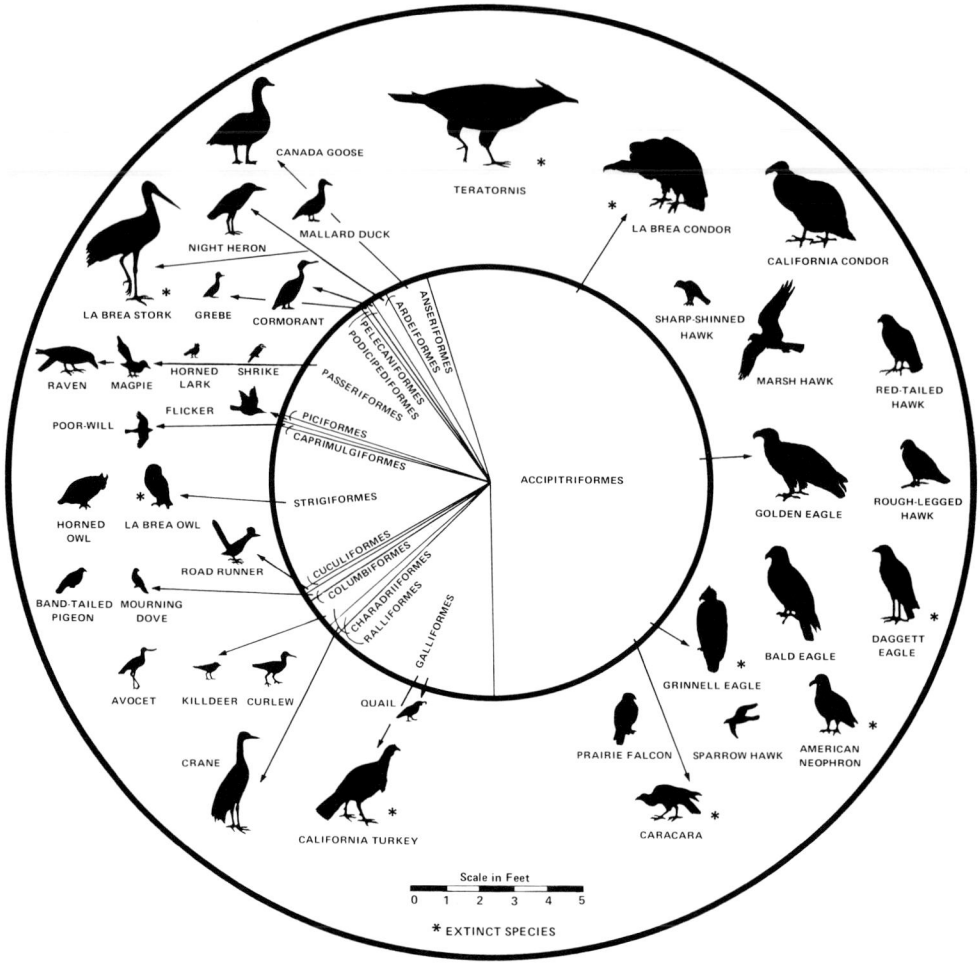

Figure 30. Diagram illustrating relative number of individuals in the avian orders occurring in the Rancho La Brea Pleistocene fauna. Note: The California Condor from the asphalt is now considered an extinct species ancestral to the living bird, and this relationship probably also existed between the Pleistocene and Recent Golden Eagles. Data from Hildegarde Howard.

mammals and other creatures trapped in the asphalt served to attract the predatory and scavenging kinds of birds from some distance away. A large number of eagles and vultures soaring and circling in the sky was probably a common sight over active asphalt seeps. Remains of raptors and of crows, ravens, and magpies are found in great numbers in the asphalt. However, it is presumed that environmental conditions as well as feeding habits were of considerable importance in contributing to a highly diversified avian population.

PODICIPEDIFORMES (Grebes)

The presence of grebes at Rancho La Brea is indicated by two specimens in the Page Museum collection, one representing a typical Grebe (*Podiceps* sp.), and the other a Pied-billed Grebe (*Podilymbus podiceps* (Linnaeus)). While grebes are known to occur elsewhere in Pleistocene deposits of western North America, notably at Fossil Lake in southern

Oregon and Lake Manix in the Mojave Desert, their infrequent occurrence in the asphalt suggests an environment in which large permanent lakes or ponds were absent. No doubt temporary bodies of water of varying size occupied the natural depressions at the Rancho La Brea locality, particularly during the wet seasons, and these might well have served to attract aquatic birds.

ARDEIFORMES (Herons and Egrets)

Although a considerable variety of forms from this group occurs in the Pleistocene asphalt, most of the species are represented by relatively few individuals. The rarer species include the Great Egret (*Casmerodius albus* (Gmelin)), Snowy Egret (*Egretta thula* (Molina)), a heron related perhaps to the Little Blue Heron (*Egretta caerula* (Linnaeus)), the Green-backed Heron (*Butorides striatus* (Linnaeus)), Black-crowned Night Heron (*Nycticorax nycticorax* (Linnaeus)), White-faced Ibis (*Plegadis chihi* (Linnaeus)), and probably the Roseate Spoonbill (*Ajaia ajaja* (Linnaeus)), each rarely represented by more than one individual. Slightly more frequent in occurrence are the Great Blue Heron (*Ardea herodias* Linnaeus) and the American Bittern (*Botaurus lentiginosus* (Montagu)). All of these birds are apparently closely related to or identical with their living representatives.

ANSERIFORMES (Waterfowl)

The occurrence of ducks and geese at Rancho La Brea is another indication of the presence of surface pools of water. While these birds are recorded in greater abundance than the grebes, their total number is still rather limited. In so far as determinations have been made from the fossil material, most of the specimens are either identical with or closely related to living species. The assemblage includes the Tundra Swan (*Cygnus columbianus* (Ord)), Canada Goose (*Branta canadensis* (Linnaeus)), Greater White-fronted Goose (*Anser albifrons* (Scopoli)), Snow Goose (*Chen caerulescens* (Pallas)), Ross' Goose (*Chen rossi* (Cassin)), Mallard (*Anas platyrhynchos* Linnaeus), Gadwall (*Anas strepera* (Linnaeus)), Green-winged Teal (*Anas crecca* (Gmelin)), Cinnamon Teal (*Anas cyanoptera)*, the Northern Shoveller (*Anas clypeata* (Linnaeus)), and two diving ducks, one similar to the Canvasback (*Aythya valisineria* (Wilson)) and another somewhat smaller form. The only extinct anserine thus far recorded from Rancho La Brea is the Brea Pigmy Goose (*Anabernicula gracilenta* (Wetmore)), which shows characters of both the ducks and the geese but is considerably smaller and more slenderlegged than any living goose. This species has been recorded from the upper Pliocene of Arizona, from the McKittrick asphalt, and similar forms from Quaternary cave deposits in Nevada and New Mexico and from Fossil Lake, Oregon (Howard, 1946, pp. 171-173).

CICONIIFORMES (Storks, Teratorns, and New World Vultures)

The common stork from the asphalt (*Ciconia maltha* Miller; Figure 31) differs from all modern species, but appears to be most closely related to *Ciconia ciconia* (Linnaeus) and *Ciconia maguari* (Gmelin). All fossil stork remains from North America are commonly assigned to one species, *C. maltha,* which is quite constant in all characteristics except size (Miller, 1932; Howard, 1941). The stork from the Pleistocene of Florida, tending to be slightly larger than the California fossil, may represent a separate subspecies, *C. m. weillsi.* The range in size of *C. maltha* includes both the Maguari and Jabiru Storks. The fossil stork has longer wings than those of either *Ciconia maguari* or *Jabiru,* and is more slender in all respects than the latter. Regarding the occurrence of the form at Rancho La Brea,

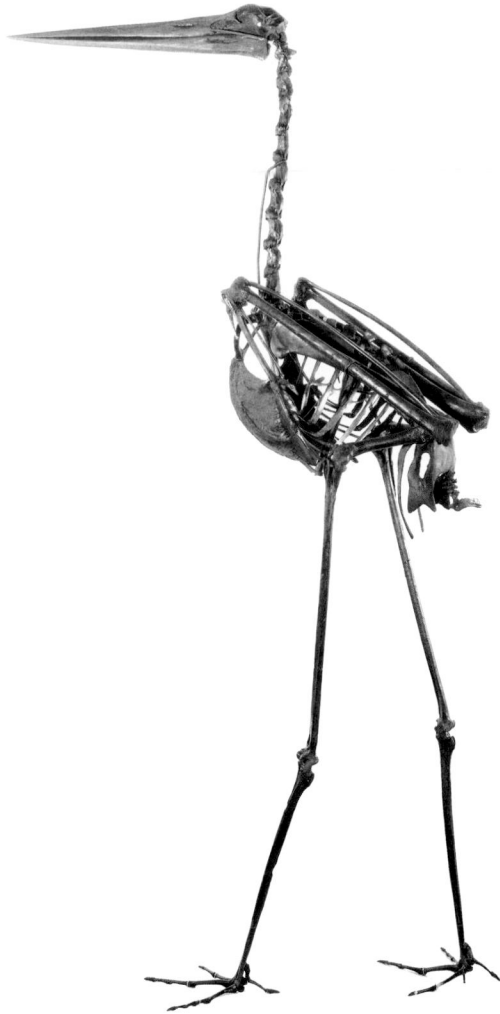

Figure 31. Composite skeleton of Asphalt Stork (*Ciconia maltha* Miller).
This mounted specimen measures 1.34 meters (4 feet 5 inches) in height.
Page Museum collection; Rancho La Brea Pleistocene. After Howard.

L. H. Miller (1925) remarked: "The fairly abundant remains of this stork in the asphalt deposits must not be interpreted as indication of a greater humidity in the region than is at present encountered there. Various writers upon the habits of storks in both Old and New Worlds speak of the plains-dwelling habit of the birds, especially during insect outbreak, such as the locust storms of Palestine and of Argentina." According to Miller, *Ciconia maltha* may have been similar in appetite and appearance to the Maguari Stork now living in the Argentine pampas.

An extinct raptorlike teratorn, *Teratornis merriami* Miller (Figure 32), was first described from Rancho La Brea, but it has since been recognized in the asphalt assemblages of Carpinteria and McKittrick, California, and in the Pleistocene of Florida and Nuevo León, Mexico. Standing about 76 cm (2.5 ft) tall, *Teratornis* possessed a wing-spread of at least

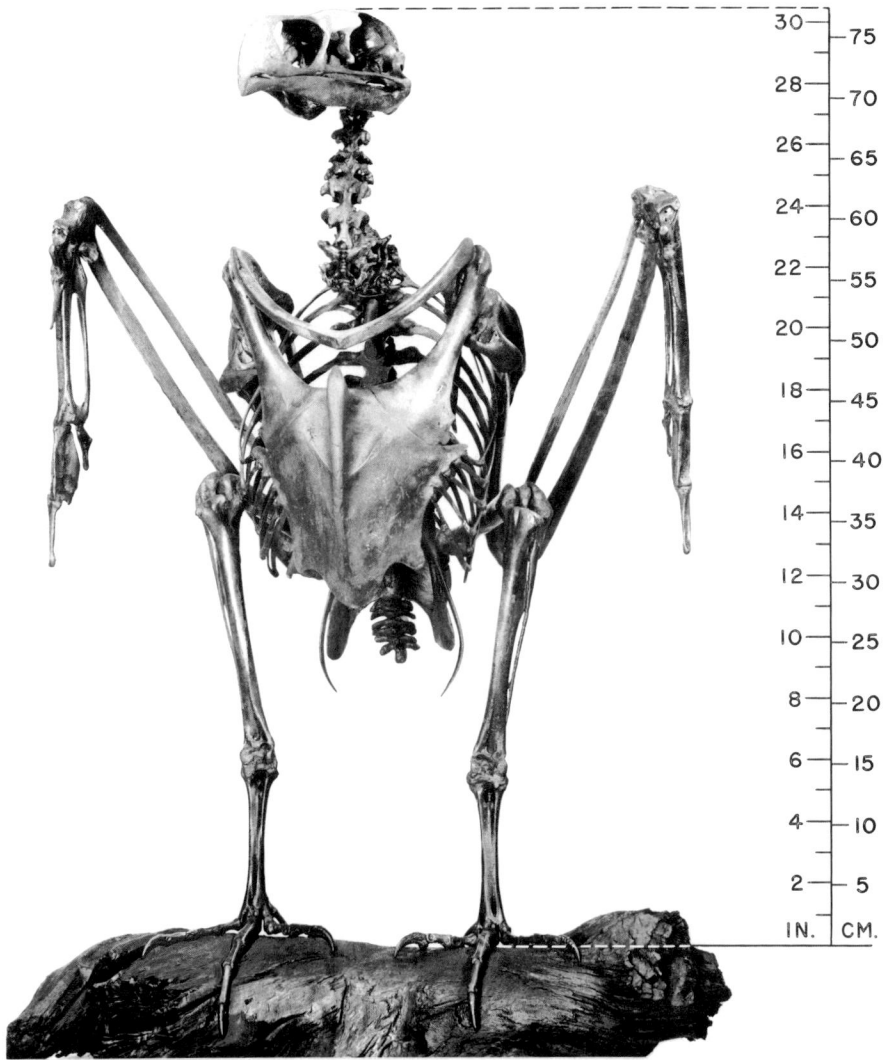

Figure 32. Composite skeleton of the great condorlike teratorn (*Teratornis merriami* Miller). The wings of this teratorn, when unfolded in life, are estimated to have measured 3.65 meters (12 feet) from tip to tip. Page Museum collection; Rancho La Brea Pleistocene.

3.5 m (12 ft) and ranks among the largest known birds of flight. In life the bird weighed possibly 14 kg (32 lb). Its skull and skeleton exhibited a curious combination of eagle- and vulture-like characters. The top of the cranium is flattened and the beak is noticeably compressed transversely, giving the head an aquiline appearance. In the structure of the skeleton, however, resemblances are seen pointing unmistakably to a kinship with the vultures. The feet are quite condorlike, but seem surprisingly small when compared with the great size of the body, wings, and head.

Several additional unique features may be mentioned with regard to the skull. The hooked beak in life was probably covered by a stout horny sheath possessing considerable strength in biting but their jaw structure was too weak to tear flesh (K. E. Campbell, pers.

comm.). The articulation of the lower jaw apparently permitted a considerable gape to the mouth. The brain-case is also relatively small.

Teratornis merriami was originally interpreted to be an extinct condorlike vulture. Recent investigation by K. E. Campbell has revealed that teratorns were, in contrast, closely related to both storks and New World vultures and were active predators rather than scavengers. Although representatives of the Family Teratornidae include *Argentavis magnificens* from South America, the largest bird known to have been capable of flight, with a wingspan of about 7 m (23 ft) (Campbell and Tonni, 1980), teratorns probably stalked their prey on the ground.

Another extinct teratorn (*Cathartornis gracilis* Miller) has been recorded from the asphalt. This bird is represented only by two tarsometatarsi which resemble those of *Teratornis* but are more slender than any bones assigned to that species.

A vulture (*Gymnogyps amplus* Miller) closely resembles in structure of skull and skeletal elements, but is slightly larger than, the existing California Condor (*Gymnogyps californianus* (Shaw)) and is represented by many individuals in the asphalt (Figure 33). Judging from the habits of this powerful scavenging bird its presence in large numbers at Rancho La Brea might well be expected. Further records of the condor in the Pleistocene have been found at Carpinteria and McKittrick (where the bird is extremely rare), in Pinellas County, Florida, and in southern Nuevo León, Mexico. Remains also occur in sub-Recent cave deposits in Nevada, Arizona, New Mexico, and Texas. *Gymnogyps amplus*, the species occurring at Rancho La Brea, was first described from the Pleistocene caves of Shasta County, California. In historic times the California Condor is known to have ranged from the Columbia River south to Baja California, but very few individuals still survive in the wild.

The Western Black Vulture (*Coragyps occidentalis* (Miller)) is closely allied to the living Black Vulture (*C. atratus* (Bechstein)) of the Midwest and southern states. The Rancho La Brea collection also includes another extinct vulture, *Breagyps clarki* (Miller), which closely parallels the living condors in size but has a distinctive long beak. This species is represented in the asphalt by several individuals. Only the Turkey Vulture (*Cathartes aura* (Linnaeus)) of the Rancho La Brea Pleistocene is still found living in the California region. Curiously enough this species is represented in the asphalt by relatively few individuals. Since Pleistocene time it has increased greatly in abundance while the California Condor has declined in numbers almost to the point of extinction.

ACCIPITRIFORMES (Diurnal Birds of Prey)

The eagle population also exhibits considerable diversity. In addition to the two species now living in North America, there are several extinct forms whose nearest relatives are found today in South America. The Golden Eagle (*Aquila chrysaetos* (Linnaeus)) has persisted from Pleistocene time into the Recent. This species occurs more abundantly at Rancho La Brea than any other bird, a census indicating in excess of 1,000 individuals in the Page Museum collection. Less numerous are the remains of the Bald Eagle (*Haliaeetus leucocephalus* (Linnaeus)), whose characters are more variable than those of the modern races of this bird.

In contrast to these forms, the morphnine eagles, with one species (*Amplibuteo woodwardi* Miller) in the Pleistocene avifauna, are now confined to Central and South America. Likewise, the crested eagles of the southern hemisphere have a related form (*Spizaetus grinnelli* (Miller)) at Rancho La Brea. Yet another type from the asphalt is the slender-limbed *Buteogallus fragilis* (Miller), which is related to a species now found in the southern part of the United States. Perhaps the most interesting of all is a long-legged eagle (*Wet-*

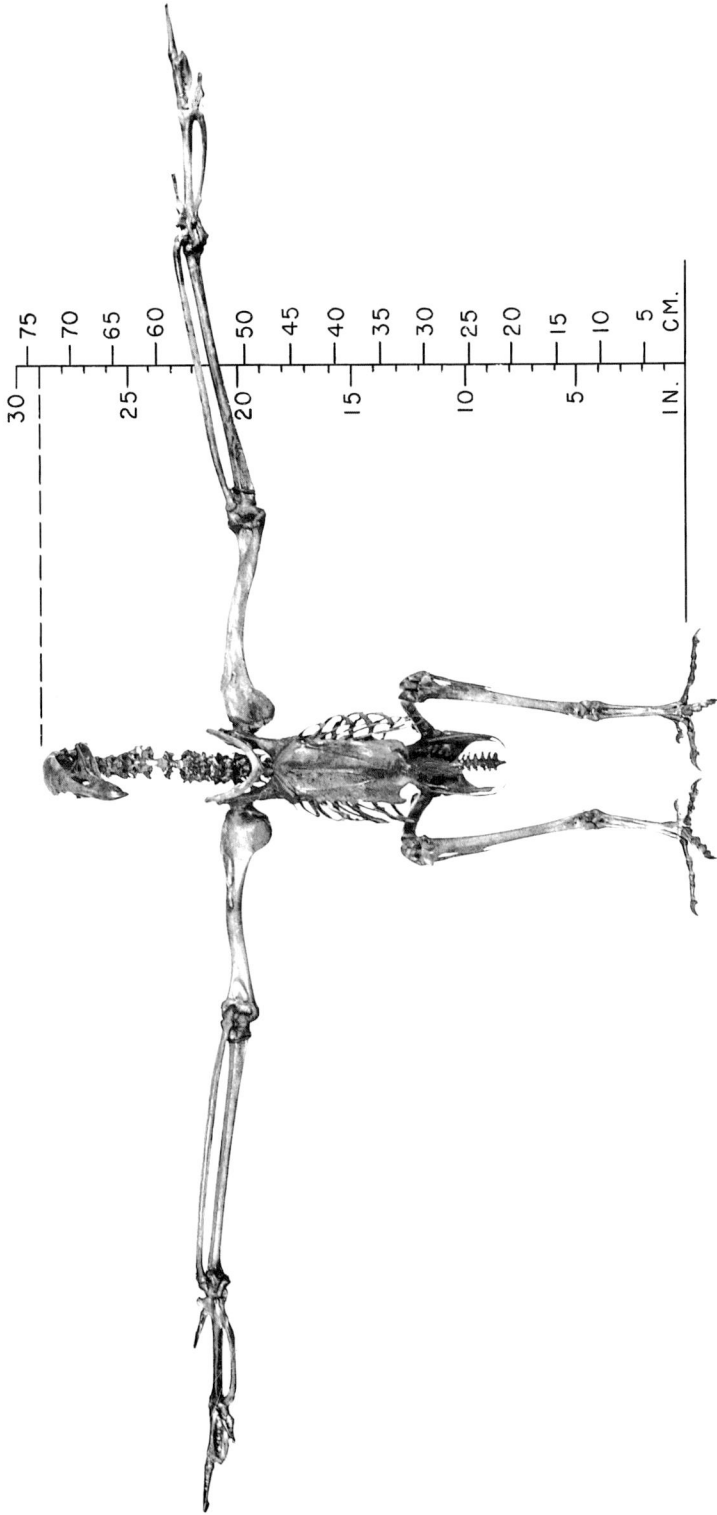

Figure 33. Skeleton of the extinct condor (*Gymnogyps amplus* Miller). Page Museum collection, Rancho La Brea Pleistocene.

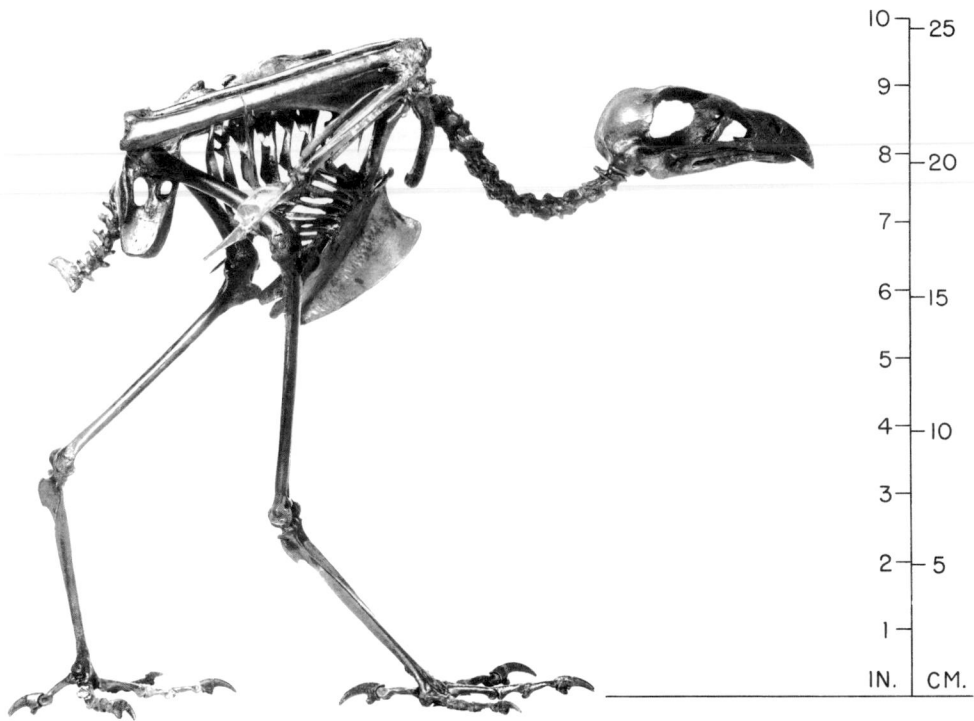

IN.	CM.
10	25
9	
8	20
7	
6	15
5	
4	10
3	
2	5
1	

Figure 34. Composite skeleton of La Brea Caracara (*Polyborus plancus prelutosus* Howard). Page Museum collection; Rancho La Brea Pleistocene. After Howard (1938).

moregyps daggetti (Miller)). Resembling superficially the living Secretary Bird of Africa and the Central and South American *Amplibuteo* in certain structural adaptations, *Wetmoregyps* possessed legs almost as long as those of the Great Blue Heron, suggesting ground habits considerably different from those of typical members of the eagle tribe. The fact that this species is more abundant in the small avifauna from Carpinteria than it is at Rancho La Brea suggests, however, that *Wetmoregyps* was a forest dweller rather than an inhabitant of open plains. It is interesting to note that this species does not occur in the McKittrick asphalt, but has been found in the Pleistocene deposits of San Josecito Cave, Mexico.

Two carrion-feeders (*Neophrontops americanus* Miller and *Neogyps errans* Miller) are related to the Old World vultures, whose representatives are not now living in America. These two species, together with the extinct Western Black Vulture (*Coragyps occidentalis* (Miller)), are also members of the Pleistocene avifauna recovered from San Josecito Cave, Nuevo León, Mexico.

Remains of a small caracara, also a carrion-feeder, were early recorded from Rancho La Brea (Figure 34). It was initially considered to be closely related to the recently extinct Guadalupe Island Caracara or to a subspecies of this caracara from the Pleistocene of San Josecito Cave. It is now, however, identified as the Crested Caracara (*Polyborus plancus prelutosus* Howard).

Many distinct kinds of the smaller birds of prey are found. Kites are known by a single specimen identical with the living Black-shouldered Kite (*Elanus caerulus* (Vieillot)). Among the hawks are numbered several species, including the Northern Harrier (*Circus cyaneus* (Linnaeus)), Northern Goshawk (*Accipiter gentilis* (Wilson)), Sharp-shinned Hawk (*Accip-*

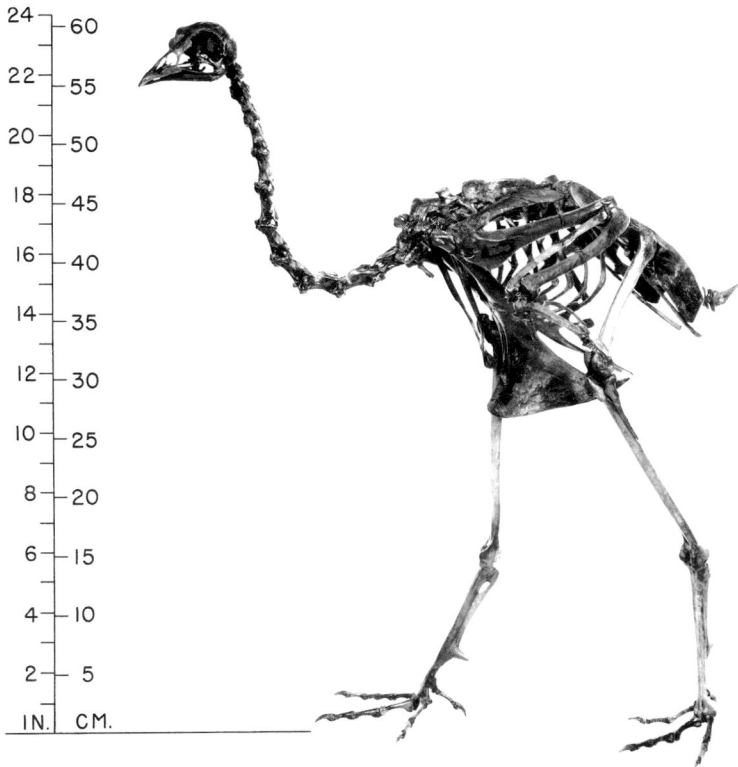

Figure 35. Composite skeleton of extinct turkey (*Meleagris californicus* (Miller)). Page Museum collection; Rancho La Brea Pleistocene.

iter striatus velox (Wilson)), Cooper's Hawk (*Accipiter cooperii* (Bonaparte)), Red-tailed Hawk (*Buteo jamaicensis* (Gmelin)), Swainson's Hawk (*Buteo swainsoni* Bonaparte), Rough-legged Hawk (*Buteo lagopus* (Brulmich)), and Ferrugineous Hawk (*Buteo regalis* (Gray)). All of the identified fossil remains of hawks have been referred to these living forms. In addition, there are a number of bones of buteonid hawks that have not yet been assigned to species.

The falcons occurring in the Rancho La Brea avifauna include the Prairie Falcon (*Falco mexicanus* Schlegel), Peregrine Falcon (*Falco peregrinus* Tunstall), Merlin (*Falco columbarius* Linnaeus), and the American Kestrel (*Falco sparverius* Linnaeus).

GALLIFORMES (Quail and Turkey)

The gallinaceous birds of the Rancho La Brea avifauna form a most interesting assemblage in which the lack of specific diversity is compensated by an extremely abundant representation of one member of this group. Only two kinds are known, of which the quail is similar to the California Quail (*Callipepla californica* (Shaw)) now inhabiting the region.

More than 700 individuals of the extinct California Turkey (*Meleagris californicus* (Miller)) are recorded in the Page Museum collection (Figure 35). In males, the shank (or tarsus) of each leg has a strong spur that is slightly heavier than in the living wild turkey. *M. californicus* approached the North and Central American wild turkey in size, and the skeleton was similarly proportioned. The nearest relative of this bird appears to be the living Ocellated Turkey of Yucatán, Mexico. The astonishing number of individuals found at Rancho La Brea, including many young birds, clearly indicates that this form may have nested locally. Its value in furnishing a potential food supply for many kinds of carnivorous animals of the Pleistocene was probably very great. The ground habits and social behavior of the extinct turkey may have made it particularly susceptible to entrapment in the asphalt seeps. It is interesting to note that while *Meleagris* occurs abundantly in the Carpinteria Pleistocene, the species is not known at all from the McKittrick asphalt. Its absence at McKittrick may have been due to a lack of suitable habitat.

GRUIFORMES (Cranes)

Representation of this group is restricted to the crane family for the most part. The forms recorded are apparently identical with living species. The Sandhill Crane (*Grus canadensis* (Linnaeus)) occurs more commonly than the Whooping Crane (*Grus americana* (Linnaeus)). A single bone of the American Coot (*Fulica americana* Gmelin) also occurs in the collection.

CHARADRIIFORMES (Shorebirds and Gulls)

These water-loving birds are not well represented individually, but several distinct types are included in the avifauna. Remains of the Killdeer (*Charadrius vociferus* (Linnaeus)), Black-bellied Plover (*Pluvialis squatarola* (Linnaeus)), Wilson's Snipe (*Gallinago gallinago delicata* (Ord)), Greater Yellow-legs (*Tringa melanoleuca* (Gmelin)), Long-billed Curlew (*Numenius americanus* Bechstein), Whimbrel (*Numenius phaeopus hudsonicus* (Latham)), Short-billed Dowitcher (*Limnodromus griseus* (Gmelin)), Marbled Godwit (*Limosa fedoa* (Linnaeus)), American Avocet (*Recurvirostra americana* Gmelin), and possibly the Black-legged Kittiwake (*Rissa tridactyla* (Linnaeus)) and Short-billed Gull (*Larus brachyrhynchus* Richardson) have been found in the Rancho La Brea deposits. These and doubtless other members of the group were probably attracted to the region by temporary ponds or perhaps by films of water that occasionally covered the surface of the asphalt seeps. The kittiwake and gull are now known to occur only in those parts of the Rancho La Brea sequence postdating 9,000 years BP.

COLUMBIFORMES (Pigeons, Doves)

Both the Band-tailed Pigeon (*Columba fasciata* Say) and the Mourning Dove (*Zenaidura macroura* (Linnaeus)) occur in the Pleistocene asphalt. The fossil remains are regarded as not differing specifically from these living types. The pigeon presumably did not frequent the region of Rancho La Brea in great numbers, for only two specimens have been definitely determined as belonging to this form. The Mourning Dove is known by at least 30 individuals, represented by 50 specimens. This great number of doves in proportion to pigeons may express the predilection of the Mourning Dove for an environment of open woods.

The Passenger Pigeon (*Ectopistes migratorius* (Linnaeus)) also occurs in the asphalt, represented by bones of three individuals. Once widely distributed over northern and eastern North America, this bird became extinct in 1914. Although now known to have lived in San Diego County, California, during the Pleistocene, there is no definite record of its existence in this area during the Recent epoch.

CUCULIFORMES (Cuckoo-like Birds)

The sole representative of this group in the asphalt is the Greater Roadrunner (*Geococcyx californianus* (Lesson)), a species which still lives in California. Its presence in the fossil assemblage may be expected since the living bird spends much of its time on the ground. Moreover, a few remains of young birds have been found.

STRIGIFORMES (Owls)

All of the species of owls occurring at Rancho La Brea, with one exception, still exist. The fossil assemblage includes the Common Barn Owl (*Tyto alba* (Scopoli)), Eastern Screech Owl (*Otus asio* (Linnaeus)), Great Horned Owl (*Bubo virginianus* (Gmelin)), Northern Pigmy Owl (*Glaucidium gnoma* (Wagler)), Burrowing Owl (*Athene cunicularia* (Molina)), Long-eared Owl (*Asio otus* (Linnaeus)), Short-eared Owl (*Asio flammeus* (Pontoppidan)), and Northern Saw-whet Owl (*Aegolius acadicus* (Gmelin)). The single extinct species is the Brea Owl (*Strix brea* Howard), related to but larger than the Barred Owl and Spotted Owl.

L. H. Miller (1925) discussed possible reasons for the relative scarcity of these nocturnal hunters in the Pleistocene avifauna of Rancho La Brea. Compared with diurnal raptors, fossil owl remains are rare both in number of specimens and of species, thereby contrasting with the balance between nocturnal and diurnal raptors today. It is possible that greater asphalt seepage during the Pleistocene may have resulted in more rapid entombment of small animals, leaving only the larger victims exposed to view and for which the small hawks and the owls would have to compete with the larger hawks. Owls were thus possibly less attracted to the asphalt trap than were the larger hawks, although today they are among the most commonly entrapped forms. An alternative explanation is that the cooling of the asphalt surface at night may have rendered the trap less dangerous to owls than to birds that hunted by day.

Howard and A. H. Miller (1939) found that the avifauna of Pit 10, associated with the human remains and of later age than the typical Pleistocene deposits at Rancho La Brea, showed a marked increase in numbers of owls, small hawks, and falcons, and concomitant decrease in numbers of the large condors and eagles, as compared with the Pleistocene assemblage. They believe that "This change in the raptors is significant particularly as it reflects the general change which has taken place in the entire fauna, that is, the disappearance of the larger Pleistocene mammals, the diminishing number of large non-raptorial birds, such as the stork and turkey, and the increasing numbers of the smaller mammals and birds." Similar changes are found in the McKittrick faunas, but differences between the older and younger assemblages are not so great as at Rancho La Brea. At Carpinteria the large raptors are less abundant than in the Pleistocene of Rancho La Brea, but in this case age does not appear to be the most important factor. It seems probable that the large raptors did not frequent this area, which was forested during the Pleistocene, preferring more open country. The relatively great abundance of the extinct California turkey, which might be expected to favor brushy or wooded regions, seems to indicate that the Carpinteria asphalt fauna approximates in age the Pleistocene assemblage at Rancho La Brea (DeMay, 1941a, b).

PICIFORMES (Woodpeckers)

Of this group, only three species, the Red-shafted Flicker (*Colaptes auratus cafer* (Gmelin)), Pileated Woodpecker (*Dryocopus pileatus* (Linnaeus)), and Lewis' Woodpecker (*Melanerpes lewisi* (Gray)), have been recorded from the Pleistocene of Rancho La Brea. Remains of flickers are not entirely unexpected in the asphalt in view of the ground feeding habits of the living birds. Since Lewis' Woodpecker is associated today with the coast live oak, it is of interest to note that remains of this tree are relatively abundant in the Pleistocene deposits. The Pileated Woodpecker, however, seems quite out of place in the Rancho La Brea fauna. It usually inhabits extensive coniferous forests, and these would not have been closely adjacent to Rancho La Brea in Pleistocene time.

PASSERIFORMES (Songbirds)

No less than 36 different kinds of passerine birds have been recognized in the fossil assemblage chiefly as a result of studies by Alden H. Miller. Of this number, 27 are identical with living species, namely the Horned Lark (*Eremophila alpestris* (Linnaeus)), Steller's Jay (*Cyanocitta stelleri* (Gmelin)), California Scrub Jay (*Aphelocoma coerulescens californica* (Vigors)), Yellow-billed Magpie (*Pica nuttalli* (Audubon)), Common Raven (*Corvus corax* Linnaeus), Chihuahuan Raven (*Corvus cryptoleucus* Couch), American Crow (*Corvus brachyrhynchos* Brehm), Northwestern Crow (*Corvus caurinus* Baird), California Thrasher (*Toxostoma redivivum* (Gambel)), Sage Thrasher (*Oreoscoptes montanus* (Townsend)), Cedar Waxwing (*Bombycilla cedrorum* Vieillot), Loggerhead Shrike (*Lanius ludovicianus* Linnaeus), Western Meadowlark (*Sturnella neglecta* Audubon), Black-headed Grosbeak (*Pheucticus melanocephalus* (Swainson)), Evening Grosbeak (*Coccothraustes vespertinus* (Cooper)), Pine Siskin (*Carduelis pinus* (Wilson)), American Goldfinch (*Carduelis tristis* (Linnaeus)), Brown Towhee (*Pipilo fuscus* Swainson), Rufous-sided Towhee (*Pipilo erythrophthalamus* (Swainson)), Vesper Sparrow (*Pooecetus gramineus* (Gmelin)), Lark Sparrow (*Chondestes grammacus* (Say)), Black-throated Sparrow (*Amphispiza bilineata* (Cassin)), Sage Sparrow (*Amphispiza belli* (Cassin)), Chipping Sparrow (*Spizella passerina* (Bechstein)), White-crowned Sparrow (*Zonotrichia leucophrys* (Forster)), Fox Sparrow (*Passerella iliaca* (Merrem)), and Song Sparrow (*Melospiza melodia* (Wilson)). An extinct species of Towhee (*Pipilo angelensis* Dawson), is also known. The type of food available during the active period of the traps frequently attracted to this locality such omnivorous birds as the ravens, the crows, and the magpies. The last are the commonest passerines recorded in the collections.

Additional forms that have been recognized include a Kingbird (*Tyrannus* sp.), Chickadee (*Parus* sp.), Bluebird (*Sialia* sp.), possibly a Yellow-headed Blackbird (*Xanthocephalus* sp.), a Redwinged Blackbird (*Agelaius* sp.), and an Oriole (*Icterus* sp.). Finally, an extinct blackbird (*Euphagus magnirostris* A. H. Miller), and an extinct icterid (*Pandanaris convexa* A. H. Miller), believed to be related to living blackbirds and cowbirds, are described from this locality. Warblers are also present but their identity still remains obscure because of the close similarity of these small species to each other, and the scarcity of distinctive skeletal elements in the Pleistocene collections.

In the light of the known habits and distribution of many of the living representatives of the passerines found as fossils at Rancho La Brea it may be presumed that the opportunities for shelter and sustenance in the country about the asphalt seeps were not unlike those indicated by the mammalian assemblage. One may reasonably infer from the variety and type of passerines in the fossil assemblage that the environment in the near vicinity of the asphalt seeps included open meadows, ground with brush cover, as well as a terrain along small stream courses with characteristic plant and tree growth.

Viewing the bird assemblages as a whole, diurnal birds of prey dominate (60 percent) followed by owls (14 percent), turkeys (13 percent), and songbirds (12 percent); lesser numbers of waterfowl and other groups are represented (Howard, 1962). Birds are more numerous in the later portion of the sequence (less than 14,000 years) and many species—particularly the water birds—are present only in the younger assemblages. Large scavenging birds, turkeys, and Whooping Cranes are present throughout the sequence but the frequency of owls and hawks increases during the younger intervals (Gust and Howard, 1991).

LOWER VERTEBRATES

REPTILES

Remains of reptiles include scattered snake vertebrae and the fragments of carapace and plastron of turtles. The latter materials apparently belong to the western pond, or mud, turtle *Clemmys marmorata* (Baird and Girard), suggesting the presence of occasional ponds or small water courses.

As the result of work by Brattstrom (1953), seven different lizards have now been identified: *Sceloporus magister* Hallowell (the desert spiny lizard), *Sceloporus occidentalis* Baird and Girard (western fence lizard), *Uta stansburiana* Baird and Girard (sideblotched lizard), *Phrynosoma coronatum* (Blainville) (coast horned lizard), *Gerrhonotus multicarinatus* (Blainville) (southern alligator lizard), *Cnemidophorus tigris* Baird and Girard (whiptailed lizard), and *Eumeces skiltoneanus* (Baird and Girard) (western skink). *Sceloporus occidentalis* is the most common lizard in the area today and in the Rancho La Brea assemblages. Except for *Sceloporus magister*, a desert margin species, all the other lizards are found today in the Los Angeles Basin (Brattstrom, 1953). Brattstrom also recognized the night lizard *Xantusia vigilis* Baird from the asphalt deposits but this identification was contested by J. M. Savage (1963, p.23).

La Duke (1983, 1991a) identified at least 14 snake species from the asphalt deposits, including *Arizona elegans* Kennicott (glossy snake), *Coluber constrictor* Linnaeus (racer), *Pituophis melanoleucus* (Daudin) (pine snake), *Lampropeltis getulus* (Linnaeus) (common king snake), *Thamnophis couchi* (Rossman and Stewart) (Sierra garter snake), *T. sirtalis* (Linnaeus) (common garter snake), *Masticophis lateralis* (Hallowell) (striped racer), *Diadophis punctatus* (Linnaeus) (ring-necked snake), and *Crotalus viridis* (Rafinesque) (western rattlesnake). All identified snake species have been reported historically from Los Angeles County and they suggest an admixture of moist riparian habitats (*Thamnophis* and *Diadophis* species) with chaparral and grassland (*Pituophus, Crotalus, Coluber,* and *Lampropeltis* species) or coastal sage associations (*Arizona, Masticophus,* and *Rhinocheilus* species) (LaDuke, 1991a).

AMPHIBIANS

Skull and skeletal materials of amphibians have also been found in the asphalt. These include the living western toad (*Bufo boreas halophilus* Baird and Girard), whose present distribution extends along the Pacific Coast from southeastern Alaska to southern California. The southwestern toad *Bufo microscaphus* Cope, the red legged true frog (*Rana aurora* Baird and Girard), and a treefrog (*Hyla* sp.) have also been documented (Brattstrom, 1953). An extinct species (*Bufo nestor* Camp), differing from the modern type in several structural characters of the skull, is also recorded. Toads were apparently as abundant at Rancho La Brea during the Pleistocene as they are at that locality today. It is of interest to record in this connection that beetles similar to those now used as food by toads occur in numbers in the asphalt.

The presence of the arboreal salamander *Aneides lugubris* (Hallowell) in the asphalt deposits has recently been documented by LaDuke (1991b). As one of the more terrestrially adapted salamanders, *A. lugubris* occurs in chaparral and live oak floral associations and still occurs today in the Los Angeles Basin.

FISH

Remains of three species of fish have been recovered from Rancho La Brea (Swift, 1979, 1989): the rainbow trout *Oncorhynchus mykiss* Walbaum, the arroyo chub *Gila orcutti* (Eigenmann and Eigenmann), and the three-spined stickleback *Gasterosteus aculeatus* Linnaeus. The type of stickleback found in the asphalt deposits is less heavily armored than most representatives of the species and both this unarmored form of stickleback and the chub are restricted to the Los Angeles Basin. In contrast, the rainbow trout is native to western North America and is found in coastal drainages from Alaska to Baja California. All the trout fossils represent individuals less than 125 mm (5 inches) long, the chub remains represent fish between 60 and 80 mm (2 and 3 in) in length, whereas the sticklebacks were 40 to 60 mm (1.5 to 2 in) long. The association of these three fish suggests the presence of permanent, slowly flowing streams.

Only four other freshwater fish are known from the Los Angeles Basin. Two are lampreys and lack bony elements suitable for preservation as fossils. The other two species—*Rhinichthys osculus* Girard (speckled dace) and *Catostomus santaanae* Snyder (Santa Ana sucker)—occur in upper stream courses in the San Gabriel Mountains; their apparent absence from the asphalt deposits supports the interpretation that streams crossing the Rancho La Brea were of local origin with headwaters in the nearby Santa Monica Mountains (Quinn, 1991).

INVERTEBRATES

Invertebrate remains were incompletely known for many years until the introduction of a new method of recovery of small specimens from the asphalt by W. Dwight Pierce of the Natural History Museum during the late 1940s. Much of the fossil insect material discussed in the ensuing paragraphs was recovered from matrix that had filled the brain cavities of the larger vertebrate skulls. Additional insect material has been recovered from the Pit 91 excavation but most has yet to be studied.

Freshwater shells, ostracod tests, and the remains of water beetles (Families Hydrophilidae, Dystiscidae, Haliplidae) and water bugs (Families Notonectidae and Corixidae) indicate the presence of more or less permanent pools of water. A few fragments of marine mollusks have been encountered in the asphalt; some of these represent accidental occurrences or intrusions, some may have been brought in by birds while others appear to have been introduced as human artifacts.

Myriapods (Diplopoda) are represented by one relatively well preserved millipede ("*Spirobolus*" *australis* Grinnell) and multitudes of fragments of several species. These are among the very few extinct invertebrate species reported from the asphaltic deposits. The Arachnida are represented by a few spiders (Araneida) and daddy longlegs (Phalangida).

The insects are by far the most abundant group, but their remains are usually fragmentary, consisting of stray head and body parts and chitinous wing covers. In addition to those mentioned above many kinds of fossil insects have been reported by Pierce, including Isoptera (termite droppings), Orthoptera (grasshoppers, crickets), Hemiptera (bugs), Homoptera (leaf hoppers), Coleoptera (beetles), Hymenoptera (wasps and ants), and Diptera (flies).

Twenty-five families of beetles have been identified from the asphalt. These include water beetles, carrion beetles, predaceous beetles, dung beetles, ground-dwelling beetles, and plant-feeding beetles including the weevils, thus giving some indication of the varying environmental conditions that prevailed. Darkling ground beetles (Tenebrionidae) are the most abundant insects preserved in California asphalt deposits; more than a dozen species are known from Rancho La Brea and most persist in the region today in coastal scrub, dry woodland and other semiarid habitats (Doyen and Miller, 1980). It seems significant that distinctive darkling beetle species currently restricted to woodland or forest habitats are not recorded from the asphalt deposits. Some of the Rancho La Brea beetle species are known only as fossils but may prove to belong to living species when their respective families have been more intensively studied (S. E. Miller, 1983).

Of the non-coleopterous forms, Jerusalem crickets and grasshoppers have been identified among the Orthoptera. The Diptera are known largely from puparia of blowflies and fleshflies. Among the Hymenoptera are ichneumons, ants, true bees, and parasitic, paper, and spider wasps. Backswimmers and water boatmen represent the Heteroptera. Many of the insects present are forms with heavy chitinous skeletons that assisted their preservation in the asphaltic deposits and most have been assigned to genera and species now living in the region. Two of the scarab beetles, *Copris pristinus* Pierce and *Onthophagous everestae* Pierce appear to be extinct. Their closest living relatives, from Mexico, are associated with mammalian dung and perhaps the terminal Pleistocene extinction of large mammalian herbivores contributed to the demise of these scarabs; alternatively, their extant representatives may persist in poorly collected regions of Mexico (S. E. Miller, 1983). Among the more unusual traces of fossil insects are two pieces of acorn with galls similar to those produced by the cynipid wasp *Callirhytis milleri* Weld (Larew, 1987).

As stated in a previous section, some of the fossil insects at Rancho La Brea suggest that the disintegration of carcasses was a comparatively slow process. Several kinds of insects have been recognized that today are characteristic of particular stages in the cycle of disintegration of organisms which follows death. Thus, blowflies and fleshflies make their appearance soon after death and subsequent stages in the postmortem period are identified by the presence of dermestid beetles (*Dermestes* sp.), of the silphid or burying and carrion beetles (*Nicrophorus* and *Heterosilpha* species) and of the histerid beetles (*Saprinus* sp.). All of the insects mentioned have been recognized as fossils in the asphalt, and these are regarded by the entomologist as evidence indicating an exposure of decaying organic matter in and about the traps for a period of at least five months.

The excavation at Pit 91 has now yielded more than 45,000 specimens of non-marine mollusks representing at least 31 species (Lamb, 1988, 1989). Seven molluskan species were associated with articulated vertebrate remains salvaged during the construction of the Page Museum (Lamb and Jefferson, 1988). The fossil mollusks are known to include 5 freshwater bivalves (*Anodonta californiensis* Lea, *Pisidium casertanum* (Poli), *P. compressum* Prime, *Musculium lacustre* (Muller), and *Musculium* cf. *M. partumeium* Say); 15 freshwater gastropods (*Valvata humeralis* Say, *Pyrgulopsis* cf. *P. californiensis* (Gregg and Taylor), *Fossaria modicella* (Say), *F. parva* (Lea), *F. (Bakerilymnea) cubensis* (Pfeiffer), *F. (B.) sonomaensis* (Hemphill), *F. (B.) bulimoides* (Lea), *F. (B.) cockerelli* (Pilsbry and Ferris), *F. (B.) cubensis* (Pfeiffer), *Stagnicola elodes* (Say), *S. proxima* (Lea), *Physella concolor* (Haldeman), *Gyraulus circumstriatus* (Tryon), *G. parvus* (Say), *Planorbella tenuis* (Dunker), and *Menetus opercularis* Gould)); and 11 terrestrial gastropods (*Pupilla hebes* (Ancey), *Vertigo occidentalis* Sterki, *Vallonia cyclophorella* Sterki, *Punctum californicum* Pilsbry, *Discus whitneyi* (Newcomb), Succineidae genus and species indeterminate, *Euconulus fulvus* (Muller), *Zonitoides arboreus* (Say), *Striatura pugetensis* (Dall), *Deroceras laeve* (Muller), and *Helminthoglypta traskii* (Newcomb).

According to Lamb (1989), the Rancho La Brea mollusks represent four aquatic habitat associations: permanent, well oxygenated, flowing water at least 30 cm (1 ft) deep; permanent, sluggish or stagnant water less than 30 cm deep; ephemeral, slow flowing to ponded water less than 30 cm deep; and mud banks of ponds and streams. Land snails represented in the fauna presently live on log or leaf litter in riparian woodland or in moist terrestrial habitats. Most of the fossil land snails comprise species that are today found in California and Arizona at elevations between 1,300 to 3,000 m (4,000 to 10,000 ft) above sea level and suggest a cooler moister climate at the time when the fossiliferous sediments of Rancho La Brea were accumulating.

Eleven species of ostracods have now been identified from the asphaltic deposits (Steinmetz, 1991). The commonest species is *Candona rawsoni* Tressler that typically occurs as a pioneer species in ephemeral ponds and streams. All the other ostracod species also belong to the Family Cyprididae and have a similar ecology to *Candona*.

PLANTS

Plants are, as a rule, excellent indicators of climate because their distributions coincide with specific ranges of temperature and humidity in particular geographic provinces. Thus, their presence at Rancho La Brea may be expected to yield interesting information regarding the climate during the period of accumulation of the Pleistocene asphalt. Many structural characters of plants are known to change only slowly in the course of geologic time, and it is therefore not surprising to find in the geologically young Rancho La Brea deposits species of plants identical with those living today in California. Interest largely centers on the association of plant types and the comparison which can be made between their past and present distributions.

Wood is the most common fossil plant material and some has been identified to species by a microscopic examination of its cellular structure. Pollen and occasional cones, seeds, and leaves also assist in making a determination of the various plants. Microscopic water plants, or diatoms, have also been recorded. It is to be regretted that the Pleistocene plant life of this region has not been more completely studied.

More than 80 species of diatoms have been recovered from sediments in the Pit 91 excavation. These include marine and brackish water forms that were restricted to asphaltic portions of the section. Large quantities of soil species occur at bone-bearing horizons, including *Hantzschia amphioxys* (Ehr.) Grun., *Pinnularia borealis* Ehr., and species of *Navicula* and *Nitzschia*. Many of the recovered diatom species are bottom-dwelling forms that typically occur in springs, marshes, and shallow alkaline lakes (Sperling, 1991).

Well over 100,000 other plant fossils are represented by pollen, leaves, seeds, cones, and wood. Although some of the 158 identified species represent plants from the local floras adjacent to the asphaltic accumulations, others may have been transported into the area by streams or flood waters. In terms of the late Pleistocene landscape, it appears that at least four plant associations were present in the area (Warter, 1976; Harris and Jefferson, 1985; Shaw and Quinn, 1986). Elements of a chaparral association were conceivably derived from the slopes of the Santa Monica Mountains to the north. Although probably not very different in overall appearance from that of today, the late Pleistocene chaparral included many plants that no longer occur in the region. The dominant plants from this association include chamise (*Adenostoma fasciculatum* Hook & Arn.), wild lilac (*Ceanothus* sp.), scrub oak (*Quercus dumosa* Nutt.), manzanita (*Arctostaphylos* sp.), walnut (*Juglans cal-*

ifornica S. Wats.), elderberry (*Sambucus mexicana* Presl. ex DC), coffeeberry (*Rhamnus californica* Esch.), and poison oak (*Toxicodendron diversilobum* Torr. & Gray). California juniper (*Juniperus californica* (Carr.) Antoine) and digger pine (*Pinus sabiniana* Dougl.) were probably located in more open, drier areas. Coast live oak (*Quercus agrifolia* Nee) probably occurred in groves on north-facing slopes, in smaller canyons, and on the lower slopes of deeper canyons. A second association, indicative of the larger, deeper, and protected canyons, includes coast redwood (*Sequoia sempervirens* (Lamb) Endl.), California bay (*Umbellularia californica* Nutt.), and dogwood (*Cornus californica* May), and evidently represents the southernmost prehistoric distribution of such an association.

The riparian (stream margin) association constitutes a mixture of plants from canyon and flood plain habitats. Sycamore (*Platanus racemosa* Nutt.), alder (*Alnus rhombifolia* Nutt.), arroyo willow (*Salix lasiolepis* Benth.), raspberry (*Rubus vitifolius* Cham. & Sch.), dogwood, poison oak, and various herbs would have been characteristic of mountain streams. Red cedar (*Juniperus* sp.), arroyo willow, sycamore, elderberry, walnut, the occasional live oak, numerous herbs, and possibly bishop pine (*Pinus muricata* Don) would have fringed the coastal plain drainages.

The coastal sage scrub association is today dominated by small, drought-tolerant, woody bushes interspersed with herbs and seasonal grasses. During the late Pleistocene, this association was dominated by coastal sage brush (*Artemesia* sp.), black sage (*Salvia* sp.), and buckwheat (*Eriogonum* sp.) with subordinate representation of saltbush (*Atriplex* sp.), tarweed (*Hemizonia fasciculata* (DC) Torre & Gray), ragweed (*Ambrosia* sp.), thistle (*Cirsium* sp.), morning glory (*Convulvulus* sp.), herbs, and grasses. Valley oaks were scattered at higher elevations on the alluvial fans; elsewhere "closed cone pine groves"— comprising Monterey pines (*Pinus radiata* Don), Monterey cypress (*Cupressus macrocarpa* Hartw.), and juniper, with an underbrush of manzanita, occurred sporadically through the landscape.

Warter (1976) reported a number of aquatic and moisture-loving plant remains from the Pit 91 excavation. These included seeds of the pondweed (*Potamogeton* sp.), bur-reed (*Sparganium eurycarpum* Engelm), and arrowhead (*Sagittaria* sp.), fruits of the horned pondweed (*Zannichella palustris* L.), at least three buttercup species (*Ranunculus aquatilis* DC Water, *R. bloomeri* S. Wats., *Ranunculus* sp.), and four species of bullrush (*Scirpus* spp.). Spikerush (*Eliocaris* sp.), willow dock (*Rumex salicifolia* Weinm.), bedstraw (*Galium trifidium* Weigand.), *Carex*, and *Verbena* species also testify to the presence of permanent water in the vicinity of Rancho La Brea.

Brattstrom (1953) concluded that from late Pleistocene to Recent there was a local transition from a moist climate typified by pine and cypress through a stage of decreasing rainfall and a vegetation of coastal live oak and California juniper, to the present day vegetation of oak woodland savanna and coastal sage scrub. Post-Pleistocene reduction in precipitation is also documented by the shrinkage and disappearance of many of the lakes formerly dotting the landscape throughout California, Nevada, and Utah during the Pleistocene (e.g., Snyder et al., 1964). The proximity of Rancho La Brea to the ocean may have ameliorated environmental fluctuations characteristic of late Pleistocene sequences at sites located farther inland. It is not yet known whether the floral associations postulated in the previous paragraphs were present in the area throughout the interval represented by the vertebrate fossils or if particular plant species or associations characterized specific intervals of time. There is evidence to suggest that Pleistocene biomes, in general, incorporated much more complex mosaics of woodland and steppe floras than might be predicted from modern ecological zonation (Guthrie, 1984).

The prevailing vegetation would, of course, have influenced the presence and distribution of the mammalian herbivores. Jefferson (1988) noted that equids predominated at coastal sites such as Costeau Pit (40,000–100,000 BP) and at the intermontane locality of McKittrick (10,000–38,000 BP), whereas camelids dominated the interior desert faunas of China Lake (10,000–42,000 BP) and Camp Cady (20,000–350,000 BP). He also found that antilocaprids and llamas formed a significant component of the intermontane fauna, edentates were relatively abundant at coastal sites, and proboscideans and bovids were present in small numbers in assemblages from all three regions. The low frequency of camelids at Rancho La Brea contrasts with desertic assemblages. The greater abundance of both bison and ground sloths at Rancho La Brea than at the other sites perhaps suggests a more wooded habitat than that represented in the intermontane basin or fringing the ocean shore. Many of the Rancho La Brea rodents are today found in wooded or chaparral habitats, although many of the larger terrestrial herbivore fossils imply more open conditions. Detailed investigation of the habitats represented by the Rancho La Brea biota will require better resolution of the temporal distribution of individual plant and animal species.

EXTINCTIONS

Of the larger mammals from Rancho La Brea, 40 percent (24 species) are now extinct. These include the numerically abundant larger carnivores and most of the more common medium- to large-sized herbivores. The youngest radiometric dates obtained from the bones of extinct Rancho La Brea mammals average about 11,000 years BP and it appears that extinction of the mammalian megafauna occurred over a relatively short time span (Marcus and Berger, 1984). The extinction of 15 percent of the bird species (21 of 138 species, mainly larger raptors) may have occurred more gradually but also took place during a relatively short interval. Possible causes of this late Pleistocene extinction event have been extensively debated during the past decade, polarizing into two major explanations—human agency (e.g., Martin, 1967, 1984) and environmental change (e.g., Graham and Lundelius, 1984; Guthrie, 1984). Proponents of the "overkill" hypothesis directly link the arrival of humans in the New World with the subsequent decline of the indigenous megafauna. Proponents of ecological hypotheses focus on the potential effect of large-scale climatic and habitat change that occurred toward the end of the Wisconsinan glaciation.

Sixty percent of the mammalian species represented by fossils at Rancho La Brea survived the extinction event. Some (such as timber wolf, cougar, pronghorn, and deer) thereafter assumed a more prominent role in the North American faunas. Some extinct species were replaced by direct descendants (bison) or near relatives (peccary). The roles of ground sloths, mammoths, mastodonts, horses, tapirs, camels and llamas were not directly filled by new ecological equivalents but by ruminants, particularly deer and pronghorn, that were present but less common during the Pleistocene (Guthrie, 1984). The terminal Pleistocene thus constituted a time of faunal turnover or extinction for the larger herbivores and the largest carnivores whose prey they comprised. Large mammals are inherently more vulnerable to environmental changes simply because they are large and require a greater expanse of primary habitat to accommodate feeding, reproductive, and defensive strategies (Guilday, 1984). Most of the large extinct herbivores shared a common, dietary-based digestive process, a fact that would seem to favor an ecological rather than anthropological cause of their demise.

Herbivorous ungulates developed two strategies to cope with the digestion of a browsing or grazing diet. Ruminants evolved a multichambered stomach, whereas monogastric caecalids enlarged the caecum and colon of the hind gut (Guthrie, 1984). Grazing specialists tend to be mammals of large size, mainly monogastric caecalids (e.g., equids, elephants, mammoths) but include some large ruminants such as bison. Forb specialists are typically small- to medium-sized ruminants, such as the pronghorn. Among the browsing species recovered from Rancho La Brea, mastodonts, ground sloths, and tapirs represent large caecalid browsers whereas deer are ruminant browsers. Of the remaining Rancho La Brea herbivores, the chambered stomach of camels and llamas is less well developed than that of other ruminants, and peccaries are monogastric omnivores.

As documented by Guthrie (1984) and others, late Pleistocene climatic changes accelerated the trend toward increased seasonality that had been initiated earlier in the Cenozoic Era. More strongly defined seasonal regimes decreased the diversity in the previously mosaic nature of plant communities and increased their zonation. The resultant shorter growing seasons for less diverse plant associations decreased the quality and quantity of food resources available to large mammalian herbivores, especially in more northerly latitudes that already had a relatively short growing season. Particularly hard hit were the mastodonts, mammoths, ground sloths, and horses—monogastric caecalids that were adapted to eating high-fiber, low-protein diets and thus able to take advantage of the mosaic vegetation of the Pleistocene grasslands. Also drastically affected were the larger carnivores whose previously plentiful food sources underwent dramatic reduction. It is interesting that the only two extinct beetles from the Rancho La Brea fauna constitute species dependant on plentiful supplies of dung for their life cycles and whose extinction might have coincided with that of the late Pleistocene ungulates.

Late Pleistocene extinctions displayed a global pattern but the phenomenon was more spectacular at higher latitudes. The Holocene relics of widely distributed Pleistocene species (lions, proboscideans, tapirs, horses, camels) are now restricted to latitudes in which annual growth seasons are comparatively long (Guthrie, 1984). It is entirely possible that the arrival of skillful human hunters in the New World, at a time of rapid, wide-sweeping environmental change, could have contributed to the ecological stress experienced by the larger mammalian herbivores by adding an additional member to the carnivore guild. In this way, man may have played a significant, though perhaps not crucial, role in the demise of the late Pleistocene megafauna. Feral horses and asses, reintroduced to North America in historic times, now survive successfully in the wild—as might close relatives of the other extinct herbivores were the continent less widely affected by human settlement.

APPENDIX

RANCHO LA BREA LOCAL BIOTA

(† = extinct, ¶ = Holocene record only,
‡ = represented by artifactual material)

FLORA

DIVISION CHRYSOPHYCOPHYTA

CLASS BACILLARIOPHYCEAE

Order Centrales

Family Coscinodiscaeae
Actinocyclus ehrenbergii var. *ralfsii* (W. Sm.) Hust.
Coscinodiscus marginatus Ehr.
C. radiatus Ehr.
Melosira italica (Ehr.) Kutz.
M. clavigera Grun.
Stephanodiscus astraea (Ehr.) Grun.

Order Pennales

Family Fragilariaceae
Fragilaria bicapitata A. Mayer var. *bicapitata*
F. virescens Ralfs var. *virescens*
F. vaucheriae (Kutz.) Peters var. *vaucheriae*
F. construens var. *binodis* (Ehr.) Grun.
Meridion circulare var. *constrictum* (ralfs) V.H.
Synedra fasciculata var. *truncata* (Grev.)
S. ulna (Nitz.) Ehr. var. *ulna*

Family Achnanthaceae
Achnanthes lanceolata (Breb.) Grun. var. *lanceolata*
Cocconeis placentula var. *lineata* (Ehr.) V.H.
C. scutellum var. *stauroneiformis* W. Smith

Family Naviculaceae
Anomoeoneis sphaerophora (Ehr.) Pfitz. var. *sphaerophora*

Caloneis bacillum (Grun.) Cl. var. *bacillum*
C. lewisii Patr. var. *lewisii*
C. limosa (Kutz.) Patr. comb. nov. var. *limosa*
C. ventricosa var. *minuta* (Grun.) Patr.
Diploneis elliptica (Kutz.) Cl. var. *elliptica*
D. didyma (Ehr.) var. *didyma*
D. pseudovalis Hust. var. *pseudovalis*
Mastogloia elliptica var. *danseii* (Thwaites) Cl.
Navicula cuspidata (Kutz.) var. *cuspidata*
N. cuspidata var. *maior* Meist.
N. halophila (Grun.) Cl. var. *halophila*
N. mutica Kutz. var. *mutica*
N. mutica Kutz. var. *cohnii* (Hilse) Grun.
N. peregrina (Ehr.) Kutz. var. *peregrina*
N. pseudoatomus Lund
N. radiosa Kutz. var. *radiosa*
N. viridula Kutz. var. *slesvicensis* (Grun.) Cl.
N. pygmaea Kutz. var. *pygmaea*
N. aurora Sov. var. *aurora*
N. laneplata (Ag.) var. *lanceolata*
N. pupula Kutz. var. *pupula*
Neidium iridis (Ehr.) Cl. var. *iridis*
Pinnularia brebissonii (Kutz.) Rabh. var. *brebissonii*
P. borealis Ehr. var. *borealis*
P. viridis (Nitz.) Ehr. var. *viridis*
P. intermedia (Lagerst.) Cl. var. *intermedia*
Stauroneis anceps cf. *S. a. linearis* (Ehr.) Hust.
S. laterostrata Hust.
S. acuta W. Sm. var. *acuta*

Family Gomphonemaceae
 Gomphonema affine Kutz. var. *affine*
 G. affine var. I (Greg.) Andrews
 G. angustatum (Kutz.) Rabh. var. *angustatum*
 G. parvulum var. *micropus* Cl.
 G. subclavatum var. *commutatum* (Grun.) A. Mayer

Family Cymbellaceae
 Amphora ovalis var. *affinis* (Kutz.) V.H. ex DeT.
 A. ovalis var. *pediculus* (Kutz.) V.H. ex DeT.
 A. venata Kutz. var. *venata*
 A. normanii Rabh. var. *mormanii*
 A. coffeiformis (Ag.) Kutz. var. *coffeiformis*
 Cymbella cistula (Ehr.) Kirchn. var. *cistula*
 C. muelleri Hust. var. *muelleri*
 C. mexicana (Ehr.) Cl. var. *mexicana*

Family Epithemiaceae
 Denticula elegans Kutz. var. *elegans*
 D. elegans var. *kittoniana* (Grun.) DeT.
 D. lauta J.W. Bail. var. *lauta*
 Epithemia emarginata Andrews var. *emarginata*
 E. turgida (Ehr.) Kutz. var. *turgida*
 E. turgida var. *westermanii* (Ehr.) Grun.
 E. adnata var. *minor* (Perag. & Herib) Patr. comb. nov.
 E. intermedia Fricke var. *intermedia*
 Rhopalodia gibba (Ehr.) O. Mull. var. *gibba*
 R. gibburula (Ehr.) O. Mull. var. *gibberula*
 R. musculus (Kutz.) O. Mull. var. *musculus*

Family Nitzchiaceae
 Hanzschia amphioxys (Ehr.) Grun.
 Nitzschia angustata var. *acuta* Grun.
 N. denticulae Grun.
 N. fonticola Grun.
 N. frustulum (Kutz.) Grun.
 N. hungarcia Grun.
 N. ignorata Krasske var. *ignorata*
 N. palea (Kutz.) W. Sm.

Taxa identified by Jon Sperling, Queens College, written comm., 1983.

DIVISION CHAROPHYTA

CLASS CHAROPHYCEAE

Order Charales
Family Characeae
 Chara sp.

DIVISION SPERMATOPHYTA

CLASS GYMNOSPERMAE

Order Pinales (=Coniferales)
Family Pinaceae
 Pinus muricata Don
 P. radiata Don
 P. sabiniana Dougl.

Family Taxodiaceae
 Sequoia sempervirens (Lamb) Endl.

Family Cupressaceae
 Cupressus forbesii Jeps.
 C. goveniana Gord.
 C. macrocarpa Hartw.
 Juniperus cf. *J. barbadensis*
 J. californica (Carr.) Antoine

CLASS ANGIOSPERMAE

Subclass Monocotyledoneae

Order Pandanales
Family Sparganiaceae
 Spariganium eurycarpum Engelm.

Order Najadales
Family Najadaceae
 Najas sp.

Family Alismataceae
 Sagittaria sp.

Family Potamogetonaceae
 Potamogeton sp.

Family Zannichelliaceae
 Zannichellia palustris L.

Order Poales (=Graminales)
Family Cyperaceae
 Scirpus sp. 1
 Scirpus sp. 2
 Scirpus sp. 3
 Scirpus sp. 4
 Heleocharis (Eleocharis) sp.
 Carex sp.

Family Poaceae (=Gramineae)
Tribe Festuceae
 Bromus sp.
 Festuca sp.

Tribe Sporoboleae
 Sporobolus sp.

Tribe Chlorideae
 Bouteloua sp.
 Hilaria sp.

Order Liliales
Family Iridaceae
Sisyrinchium bellum S. Wats.

CLASS ANGIOSPERMAE
Subclass Dicotyledoneae

Order Juglandales
Family Juglandaceae
Juglans californica S. Wats.

Order Salicales
Family Salicaceae
Salix lasiolepis Benth.

Order Utricales
Family Ulmaceae
Celtis reticulata Torr.

Order Fagales
Family Betulaceae
Alnus rhombifolia Nutt.

Family Fagaceae
Quercus agrifolia Nee
Q. dumosa Nutt.
Q. lobata Nee

Order Polygonales
Family Polygonaceae
Eriogonum sp.
Rumex salicifolia Weinm.

Order Chenopodales
Family Chenopodiaceae
Chenopodium sp.
Atriplex sp.
Eurotia lanata (Pursh) Moc.

Family Amaranthaceae
Amaranthus sp.

Order Caryophyllales
Family Portulacaceae
Calandrinia ciliata var. *menziesii* (Hook.),
J. F. McBride
Montia spathulata (Dougl.) Howell

Order Ranunculales
Family Ranunculaceae
Ranunculus aquatilis DC Water
R. bloomeri S. Wats.
Ranunculus sp.

Family Berberidaceae
Berberis sp.

Family Lauraceae
Umbellularia californica Nutt.

Order Rosales
Family Platanaceae
Platanus racemosa Nutt.

Family Rosaceae
Adenostoma fasciculatum Hook. & Arn.
Rubus vitifolius Cham. and Sch.

Order Sapindales
Family Anacardiaceae
Toxicodendron diversiloba Torr. and Gray

Family Aceraceae
Acer negundo L.

Family Balsaminaceae
Balsamorhiza deltiodea Nutt.

Order Rhamnales
Family Rhamnaceae
Rhamnus californica Esch.
Ceanothus sp.

Order Malvales
Family Malvaceae
Sphaeralcea sp.

Order Apiales (=Umbellales)
Family Apiaceae (=Umbellifereae)
Berula erecta (Huds.) Coville
Oenanthe sarmentosa Presl.

Family Cornaceae
Comus californica C. A. May

Order Ericales
Family Ericaceae
Xylococcus bicolor Nutt.
Arctostaphylos glauca Lindl.
A. insularis Green
A. morroensis Weisl. and Schr.
A. pechoensis Dudley
A. pungens H. B. K.
A. tomentosa (Prsch.) Lindl.
A. viscida Parry

Order Polemoniales
Family Convolvulaceae
Convolvulus sp.

Family Lamiaceae (=Labiatae)
Salvia sp.

Family Scrophulariacaea
  Antirrhinum nuttallianum Benth.
  Orthocarpus purpurascens Benth.

Family Verbenaceae
  Verbena sp. 1
  Verbena sp. 2

Order Rubiales
Family Rubiaceae
  Galium trifidum

Family Caprifoliaceae
  Sambucus mexicana Presl ex DC.

Order Campanulales
Family Asteraceae (=Compositae)
  Ambrosia sp.
  Madia sp.
  Hemizonia fasciculata (DC.) Torre and Gray
  Xanthium strumarium var. *glabratum* (DC.)
    Cronquist (=*X. s. calvum*)
  Calycadenia tenella (Nutt.) Torr. and Gray
  Artemisia sp.
  Cirsium sp.

Taxa from Frost, 1927, Stock, 1956, Templeton, 1964, Warter 1976, and Akersten, W. A., T. M. Foppe, and G. T. Jefferson, 1987.

FAUNA
PHYLUM ARTHROPODA
CLASS ARACHNIDA
Order Scorpionida
Family Vejovidae?
  gen. & sp. indet.

Order Araneida
Family Clubionidae?
  gen. & sp. indet.

Family Thomisidae?
  gen. & sp. indet.

Family Salticidae?
  gen. & sp. indet.

Family Lycosidae?
  gen. & sp. indet.

Order Phalangida
Family indet.
  gen. & sp. indet.

CLASS CRUSTACEA
Order Ostracoda
Family Cyprididae
  Candona cf. *C. acutula* Sars, 1924
  C. candida (Müller, 1776)
  Candona cf. *C. rawsoni* Tressler, 1957
  Candona sp.
  Cyclocypris cf. *C. laevis* (Müller, 1785)
  Cypridopsis vidua (Müller, 1776)
  Cyprinotus glaucus
  (?) *Cypris* sp.
  Eucypris (?) sp.
  Limnocythere sp. cf. *L. paraornata*
  Potamocypris grandulosa

Order Isopoda
Family Cytheridae
  Armadillidium vulgare (Latreille), 1804

CLASS DIPLOPODA
Order Julida
Family Julidae
  Julus occidentalis Grinnell, 1908 †
  J. cavicola Grinnell, 1908 †

Order Spirobolida
Family Spirobolidae
  "*Spirobolus*" *australis* Grinnell, 1908 †
  "*Spirobolus*" sp.

CLASS CHILOPODA
Unidentified centipedes

CLASS INSECTA
Order Orthoptera
Family Acrididae
  gen. & sp. indet.

Family Stenopelmatidae
  Stenopelmatus sp.

Order Dermaptera
Family Carcinophoridae
  Euboriella annulipes (Lucas, 1847) (contamination?)

Order Isoptera
Family indet.
  gen. & sp. indet.

Order Heteroptera
Family Corixidae
  Hemipleura sp.

Family Notonectidae
  Notonecta sp.

Family Belostomatidae
  Lethocerus americanus (Leidy, 1847)

Family Gerridae
  Gerris sp.

Family Coreidae
  Catorhintha sp.

Family Reduviidae
  Rasahus sp. cf. *R. bigguttatus* (Say, 1832)

Family Scutelleridae?
  gen. & sp. indet.

Order Homoptera
Family Membracidae
  gen. & sp. indet.

Family Cicadellidae
  gen. & sp. indet.

Order Coleoptera

Suborder Adephaga
Family Cicindelidae
  Cicindela haemorrhagica LeConte, 1851
  C. oregona LeConte, 1857

Family Carabidae
  Amara insignis Dejean, 1831
  Agonum maculicolle (Dejean, 1828)
  Agonum sp.
  cf. *Bembidion* sp.
  Calosoma semilaeve LeConte, 1849
  Calosoma near *C. cancellatum* Eschscholtz, 1829
  Calosoma sp.
  Dicheirus sp.
  Elaphrus californicus Mannerheim, 1843
  Elaphrus sp. cf. *E. finitimus* Casey, 1920
  Elaphrus sp.
  Platynus sp. cf. *P. funebris* LeConte, 1854
  Pterostichus sp.
  cf. *Trechus* sp.
  gen. & sp. indet.

Family Haliplidae
  Peltodytes sp.
  gen. & sp. indet.

Family Dytiscidae
  Acilius sp.
  Colymbetes strigatus " *celaenus*" Pierce

Colymbetes near *C. strigatus* LeConte, 1851
Colymbetes sp.
Cybister ellipticus LeConte, 1851
Cybister sp.
Deronectes sp.
Dytiscus sp.
Thermonectes sp.

Family Gyrinidae
  Gyrinus sp.

Suborder Myxophaga
Family Sphaeriidae
  gen. & sp. indet.

Suborder Polyphaga
Family Histeridae
  Hololepta vicina LeConte, 1851
  Saprinus sp.
  gen. & sp. indet.

Family Hydrophilidae
  Helophorus sp.
  Hydrous triangularis (Say, 1823)
  Hydrous sp.
  Tropisternis sp.
  gen. sp. indet.

Family Limnebiidae?
  gen. & sp. indet.

Family Leiodidae
  cf. *Agthidium* sp.

Family Silphidae
  Heterosilpha ramosa (Say, 1823)
  Heterosilpha sp.
  Nicrophorus guttula Motschoulsky, 1845
  N. marginatus Fabricius, 1801
  N. nigrita Mannerheim
  Nicrophorus sp.
  Thanatophilus lapponicus (Herbst, 1793)

Family Staphylinidae
  gen. sp. indet.

Family Dermestidae
  Dermestes sp.

Family Tenebrionidae
  Apsena laticornis Casey, 1851
  A. pubescens (LeConte, 1851)
  A. rufipes Eschscholtz, 1829
  Apsena sp.
  Coniontis abdominalis LeConte, 1859
  C. elliptica Casey, 1884

C. *remnans* Pierce, 1954
C. *lamentabilis* Blaisdell, 1924
C *robusta* Horn, 1870
C. *rugosa* Casey, 1908
Coniontis sp.
Cratidus osculans LeConte
Cratidus sp
Eleodes acuticaudus LeConte, 1851
E. *acuticaudus punctata* Blaisdell, 1909
E. *borealis* Blaisdell, 1909
E. *consobrina* LeConte, 1851
E. *dentipes* Eschscholtz, 1829
E. *dentipes elegans* Casey, 1890
E. *dentipes pertenuis* Blaisdell, 1909
E. *distans* Blaisdell, 1909
E. *gigantea meridionalis* Blaisdell, 1918
E. *grandicollis grandicollis* Mannerheim, 1843
E. *grandicollis* Mannerheim, 1843, subspp.
E. *laticollis* LeConte, 1851
E. *omissus* LeConte, 1858
E. *omissus omissus* LeConte, 1858
E. *o. pygmaea* Blaisdell, 1909
E. *osculans* (LeConte)
E. *punctata* Blaisdell, 1909
Eleodes sp.
Nyctoporis carinata LeConte, 1851
Nyctoporis sp.

Family Zopheridae
Noserus plicatus LeConte, 1859
Phloeodes pustulosus LeConte, 1859
Phloeodes sp.

Family Scarabaeidae
Canthon (Boreocathon) simplex LeConte, 1857
C. *(Boreocathon) praticola* LeConte, 1859
Copris pristinus Pierce, 1947
Copris sp.
Onthophagus everestae Pierce, 1947
Onthophagus sp.
Phanaeus labreae (Pierce, 1947)
Phileurus illatus LeConte, 1854
Serica kanakoffi Pierce, 1947
Trox gemmulatus Horn, 1874
T. *suberosus* (Fabricius, 1775)

Family Heteroceridae
Heterocerus sp.

Family Dryopidae
gen. & sp. indet.

Family Elateridae
Acolus sp.
Anchastus cinereipennis (Eschscholtz, 1829)

Aplastus sp.
Cardiophorus sp.
Dalopius sp. 1.
Dalopius sp. 2.
Dalopius sp. 3.
Limonius sp.
Melanotus sp.
Melanotus (?) sp.
gen. & sp. indet.

Family Coccinellidae
Coccinella californica Mannerheim, 1843

Family Anthicidae
Notoxus sparsus LeConte, 1859

Family Chrysomelidae
gen. & sp. indet.

Family Cerambycidae
gen. & sp. indet.

Family Curculionidae
Apion sp.
Dinocleus sp.
gen. & sp. indet.

Family Scolytidae
Gnathotrichus sp.

Order Lepidoptera
Family indet.
gen. & sp. indet.

Order Diptera
Family Bibionidae
Dilophus sp.

Family Calliphoridae
Cochliomyia macellaria (Fabricius, 1775)
Cochliomyia sp.

Order Hymenoptera
Family Chalcididae
gen. & sp. indet.

Family Bethylidae?
gen. & sp. indet.

Family Ichneumonidae
gen. & sp. indet.

Family Formicidae
Camponotus sp.
Camponotus near C. *quericola* M. Smith, 1953
Camponotus near C. *vicinus* Mayr, 1870

Formica moki Wheeler, 1906
Formica near *F. moki* Wheeler, 1906
Formica sp.
Pheidole sp.
Solenopsis sp.
Messor andrei (Mayr, 1886)
Messor sp.

Family Vespidae
    gen. & sp. indet.

Family Pompilidae
    gen. & sp. indet.

Family Cynipidae
    ?Callirhytis sp.

Superfamily Apoidea
    gen. & sp. indet.

Taxonomy and identifications follow Borror, et al., 1976, and Miller, 1983, with assistance from R. Snelling.

PHYLUM MOLLUSCA

CLASS BIVALVIA

Subclass Paleoheterodonta

Order Unionoida
Family Unionidae
    Anodonta californiensis Lea, 1852

Subclass Heterodonta

Order Veneroida
Family Cardiidae
    Trachycardium quadragenarium (Conrad, 1837) ‡
    Laevicardium elatum (Sowerby, 1835) ‡

Family Donacidae
    Donax gouldii Dall, 1921

Family Lucinidae
    Epilucina californica (Conrad, 1837) ‡

Family Mactridae
    Tresus nuttalli (Conrad, 1837) ‡

Family Pisidiidae (=Sphaeriidae)
    Pisidium casertanum (Poli, 1795)
    P. compressum Prime, 1852
    Musculium lacustre (Müller, 1774)
    Musculium cf. *M. partumeium* (Say, 1822)

Family Veneridae
    Transennella sp.

Tivela stultorum (Mawe, 1823) ‡
Saxidomus nuttalli Conrad, 1837 ‡
Chione californiensis (Broderip, 1835) ‡
C. undatella (Sowerby, 1835) ‡

Order Myoida
Family Corbulidae
    Corbula luteola Carpenter, 1864

Subclass Pteriomorphia

Order Pterioida
Family Ostreidae
    Ostrea lurida Carpenter, 1864 ‡

Family Pectinidae
    Argopecten circularus aequisulcatus (Carpenter, 1864) ‡
    gen. & sp. indet.

CLASS GASTROPODA

Subclass Prosobranchia

Order Archeogastropoda
Family Acmaeidae
    Notacmaea insessa (Hinds, 1843)

Family Haliotidae
    Haliotis rufescens Swainson, 1822 ‡

Family Trochidae
    Lirularia optabilis (Carpenter, 1864) †

Family Turbinidae
    Astraea undosa (Wood, 1828) ‡

Order Mesogastropoda
Family Hydrobiidae
    Pyrogulopsis californiensis (Gregg and Taylor, 1965)

Family Lacunidae
    cf. *Lacuna* sp.

Family Naticidae
    Lunatia lewisii (Gould, 1847) ‡

Family Rissoidae
    Alvinia compacta (Carpenter, 1864)

Family Valvatidae
    Valvata humeralis Say, 1829

Order Neogastropoda
Family Columbellidae
    Mitrella carinata (Hinds, 1844)
    Nassarina penicillata (Carpenter, 1864)

Family Olividae
  Olivella baetica (Carpenter, 1864)
  Olivella cf. *O. biplicata* (Sowerby, 1825)

Subclass Pulmonata

Order Basommatophora
Family Lymnaeidae
  Fossaria modicella (Say, 1825)
  F. parva (Lea, 1841)
  Fossaria (Bakerilymnaea) bulimoides (Lea, 1841)
  F. (B.) cockerelli (Pilsbry and Ferris, 1906)
  F. (B.) cubensis (Pfeiffer, 1839)
  F. (B.) sonomaensis (Hemphill in Pilsbry and Ferris, 1906)
  Stagnicola elodes (Say, 1821)
  S. proxima (Lea, 1856)

Family Physidae
  Physella concolor (Haldeman, 1841)

Family Planorbidae
  Gyraulus circumstriatus (Tryon, 1866)
  G. parvus (Say, 1824)
  Planorbella tenuis (Dunker, 1850)
  Menetus opercularis (Gould, 1847)

Order Stylommatophora
Family Discidae
  Discus whitneyi (Newcomb, 1864)

Family Helminthoglyptidae
  Helminthoglypta traskii (Newcomb, 1861)

Family Limacidae
  Deroceras laeve (Müller, 1774)

Family Punctidae
  Punctum californicum Pilsbry, 1898

Family Pupillidae
  Pupilla hebes (Ancey, 1881)
  Vertigo occidentalis Sterki, 1907

Family Succineidae
  gen. & sp. indet.

Family Urocoptidae
  Holospira sp.

Family Valloniidae
  Vallonia cyclophorella Sterki, 1892

Family Zonitidae
  Zonitoides arboreus (Say, 1816)
  Striatura pugetensis (Dall, 1895)

Taxonomy of freshwater/terrestrial species from identifications by A. G. Smith, California Academy of Science, written comm. 1972; W. 0. Gregg 1944; C. C. Coney pers. comm. 1987; R. V. Lamb pers. comm. 1987; D. W. Taylor pers. comm. 1986; B. Roth pers. comm. 1987; marine/artifactual species from R. L. Reynolds pers. comm. 1983; R. V. Lamb pers. comm. 1987; J. McLean pers. comm. 1987; D. Lindberg pers. comm. 1987.

PHYLUM CHORDATA

CLASS OSTEICHTHYES

Order Salmoniformes
Family Salmonidae
  Oncorhynchus mykiss Walbaum 1792

Order Cypriniformes
Family Cyprinidae
  Gila orcutti (Eigenmann and Eigenmann, 1890)

Order Gasterosteiformes
Family Gasterosteidae
  Gasterosteus aculeatus Linnaeus, 1758

Taxonomy and identifications by Dr. C. Swift, Los Angeles County Museum of Natural History, pers. comm., 1983.

CLASS AMPHIBIA

Order Urodela
Family Plethodontidae
  Aneides lugubris (Hallowell, 1849)

Order Anura
Family Bufonidae
  Bufo nestor (Camp, 1917) †
  B. boreas Baird and Girard, 1852
  B. microscaphus Cope, 1867

Family Hylidae
  Hyla sp.

Family Ranidae
  Rana aurora Baird and Girard, 1852

CLASS REPTILIA

Order Chelonia
Family Emydidae
  Clemmys marmorata (Baird and Girard, 1852)

Order Squamata
Family Anguidae
  Gerrhonotus multicarinatus (Blainville, 1835)

Family Iguanidae
 Phrynosoma coronatum (Blainville, 1835)
 Sceloporus magister Hallowell, 1854
 S. occidentalis Baird and Girard, 1852
 Uta stansburiana Baird and Girard, 1852

Family Scincidae
 Eumeces skiltonianus (Baird and Girard, 1852)

Family Teiidae
 Cnemidophorus tigris Baird and Girard, 1852

Family Colubridae
 Arizona elegans Kennicott, 1859
 Coluber constrictor Linnaeus, 1766
 C. constrictor mormon Baird and Girard, 1852
 Diadophis punctatus (Linnaeus, 1766)
 Hypsiglena torquata (Gunther, 1860)
 Lampropeltis getulus (Linnaeus, 1766)
 Lampropeltis sp.
 Masticophis lateralis (Hallowell, 1853)
 Pituophis melanoleucus (Daudin, 1803)
 Rhinocheilus lecontei Baird and Girard, 1853.
 Tantilla sp.
 Thamnophis cf. *T. couchi* (Rossman and Stewart, 1987)
 T. sirtalis (Linnaeus, 1758)
 Thamnophis sp.

Family Viperidae
 Crotalus viridis (Rafinesque, 1818)
 Crotalus sp.

Reptile and Amphibian taxa from Brattstrom, 1953, La Duke, 1983, and J. A. Holman, Michigan State University, pers. comm., 1977.

CLASS AVES

Order Podicipediformes
Family Podicipedidae
 Podilymbus podiceps (Linnaeus, 1758)
 Podiceps sp.

Order Pelecaniformes
Family Phalacrocoracidae
 Phalacrocorax sp.

Order Ardeiformes
Family Ardeidae
 Ardea herodias Linnaeus, 1758
 Botaurus lentiginosus (Rackett, 1813)
 Butorides striatus (Linnaeus, 1758)
 Casmerodius albus Linnaeus, 1758
 Egretta thula (Molina, 1782)
 E. caerulea (Linnaeus, 1758)
 Nycticorax nycticorax (Linnaeus, 1758)

Family Threskiornithidae (Plataleidae)
 Ajaia ajaja (Linnaeus, 1758)
 Plegadis chihi (Vieillot, 1817)

Order Anseriformes
Family Anatidae
 Anas platyrhynchos Linnaeus, 1758
 A. strepera Linnaeus, 1758
 A. crecca Linnaeus, 1758
 A. cyanoptera (?) Vieillot, 1816
 A. clypeata Linnaeus, 1758
 Anser albifrons (Scopoli, 1769)
 Aythya valisineria Wilson, 1814
 Anabernicula gracilenta (Ross, 1935) †
 Branta canadensis (Linnaeus, 1758)
 Branta cf. *B. bernicla* (Linnaeus, 1758)
 Chen caerulescens (Linnaeus, 1758)
 C. rossi (Cassin 1861)
 Cygnus columbianus (Ord, 1815)

Order Ciconiiformes
Family Ciconiidae
 Ciconia maltha Miller, 1910 †
 Mycteria wetmorei Howard, 1935 †

Family Teratornithidae
 Cathartornis gracilis Miller, 1910 †
 Teratornis merriami Miller, 1909 †

Family Vulturidae (Cathartidae)
 Breagyps clarki (Miller), 1910 †
 Coragyps occidentalis (Miller), 1909 †
 Cathartes aura (Linnaeus, 1758)
 Gymnogyps amplus Miller, 1911 †

Order Accipitriformes
Family Accipitridae
 Accipiter gentilis (Linnaeus, 1758)
 A. striatus velox (Vieillot, 1807)
 A. cooperii (Bonaparte, 1828)
 Aquila chrysaetos (Linnaeus, 1758)
 Buteo jamaicensis (Gmelin, 1788)
 B. swainsoni Bonaparte, 1838
 B. lagopus (Pontoppidan, 1763)
 B. regalis (Gray, 1844)
 Buteo sp.
 Buteogallus fragilis (Miller, 1911) †
 Circus cyaneus (Linnaeus, 1766)
 Elanus caeruleus (Desfontaines, 1789)
 Haliaeetus leucocephalus (Linnaeus, 1766)
 Amplibuteo woodwardi (Miller, 1911) †
 Wetmoregyps daggetti (Miller, 1915) †
 Spizaetus grinnelli (Miller, 1911) †
 Neogyps errans (Miller, 1916) †
 Neophrontops americanus Miller, 1916†

Family Falconidae
 Falco columbarius Linnaeus, 1758

F. *mexicanus* Schlegel, 1851
F. *peregrinus* Tunstall, 1771
F. *sparverius* Linnaeus, 1758
Falco sp.
Polyborus plancus (Miller, 1777)

Order Galliformes
Family Phasianidae
Meleagris californica (Miller, 1909) †
Callipepla californica (Shaw, 1798)

Order Gruiformes
Family Rallidae
Fulica americana (Gmelin, 1789)

Family Gruidae
Grus canadensis (Linnaeus, 1758)
G. americana (Linnaeus, 1758)

Order Charadriiformes
Family Charadriidae
Charadrius vociferus Linnaeus, 1758
Pluvialis squatarola (Linnaeus, 1758)

Family Recurvirostridae
Recurvirostra americana Gmelin, 1789

Family Scolopacidae
Calidris alba (Pallas, 1764)
C. alpina (Linnaeus, 1758)
Catoptrophorus semipalmatus (Gmelin, 1789)
Gallinago gallinago delicata (Linnaeus, 1758)
Limnodromus griseus (Gmelin, 1789)
Limosa fedoa (Linnaeus, 1758)
Numenius americanus (Bechstein, 1812)
N. phaeopus hudsonicus (Linnaeus, 1758)
Tringa melanoleuca (Gmelin, 1789)
Phalaropus fulicarius (Linnaeus, 1758)

Family Laridae
Larus canus Linnaeus, 1758 ¶
Rissa tridactyla (Linnaeus, 1758) ¶

Order Columbiformes
Family Columbidae
Columba fasciata Say, 1823
Ectopistes migratorius (Linnaeus, 1766) †
Zenaida macroura (Linnaeus, 1758)

Order Cuculiformes
Family Cuculidae
Geococcyx californianus (Lesson, 1829)

Order Strigiformes
Family Tytonidae
Tyto alba (Scopoli, 1769)

Family Strigidae
Aegolius acadicus (Gmelin, 1788)
Asio flammeus (Pontoppidan, 1763)
Athene cunicularia (Molina, 1782)
Bubo virginianus (Gmelin, 1788)
Glaucidium gnoma Wagler, 1832
Otus asio (Linnaeus, 1758)
Strix brea Howard, 1933 †

Order Caprimulgiformes
Family Caprimulgidae
Phalaenoptilus nuttallii (Audubon, 1844)

Order Piciformes
Family Picidae
Colaptes auratus cafer (Linnaeus, 1758)
Dryocopus pileatus (Linnaeus, 1758)
Melanerpes lewisi (Gray, 1849)
Picoides sp.
Sphyrapicus sp.

Order Passeriformes
Family Tyrannidae
Tyrannus vociferans Swainson, 1826

Family Alaudidae
Ermophila alpestris (Linnaeus, 1758)

Family Corvidae
Aphelocoma coerulescens (Bosc, 1795)
A. coerulescens californica
Corvus corax Linnaeus, 1758
C. cryptoleucus Couch, 1854
C. brachyrhynchos Brehm, 1822
C. caurinus Baird, 1858
Cyanocitta stelleri (Gmelin, 1788)
Nucifraga columbiana (Wilson, 1811)
Pica nuttalli (Audubon, 1837)

Family Paridae
Parus sp. cf. *P. gambeli* Ridgeway, 1866

Family Muscicapidae
Sialia sp. cf. *S. mexicana* Swainson, 1832
Turdus migratorius Linnaeus, 1766

Family Mimidae
Oreoscoptes montanus (Townsend, 1837)
Toxostoma redivivum (Gambel, 1845)

Family Bombycillidae
Bombycilla cedrorum (Vieillot, 1808)

Family Laniidae
Lanius ludovicianus Linnaeus, 1766

Family Emberizidae
Parulinae gen. sp. indet.
Pheucticus melanocephalus (Swainson, 1827)
Amphispiza bilineata (Cassin, 1850)
A. belli (Cassin, 1850)
Chondestes grammacus (Say, 1823)
Melospiza melodia (Wilson, 1810)
Passerella iliaca (Merrem, 1786)
Pipilo erythrophthalamus (Linnaeus, 1758)
P. fuscus Swainson, 1827
P. angelensis Dawson, 1948 †
Pooecetes gramineus (Gmelin, 1789)
Spizella passerina (Bechstein, 1798)
Spizella sp.
Zonotrichia leucophrys (Forster, 1772)
Agelaius sp. cf. *A. phoeniceus californicus* (?)
Euphagus magnirostris Miller, 1929 †
Icterus spp.
Molothrus ater (Boddaert, 1783)
Sturnella neglecta Audabon, 1844
Xanthocephalus sp.
Pandanaris convexa †

Family Fringillidae
Carduelis pinus (Wilson, 1810)
C. tristis (Linnaeus, 1758)
Coccothraustes vespertinus (Cooper, 1825)
Fringillidae gen. and spp. indet.

Identifications principally from H. Howard (see literature cited). Taxonomy follows the American Ornithologists Union Check List (1957; 1983), except where modified as per K. E. Campbell, Los Angeles County Museum of Natural History, pers. comm., 1982.

CLASS MAMMALIA
Order Insectivora
Family Soricidae
Sorex ornatus Merriam, 1895
Notiosorex crawfordi (Coues, 1877)

Family Talpidae
Scapanus latimanus (Bachman, 1842)

Order Primates
Family Hominidae
Homo sapiens Linnaeus, 1758 ¶

Order Chiroptera
Family Vespertilionidae
Lasiurus cinereus (Palisot de Deauvois, 1796)
Antrozous pallidus (Le Conte, 1856)

Order Edentata
Family Megalonychidae
Megalonyx jeffersonii (Desmarest, 1822) †

Family Megatheridae
Nothrotheriops shastensis (Sinclair, 1905) †

Family Mylodontidae
Glossotherium harlani (Owen, 1840) †

Order Lagomorpha
Family Leporidae
Sylvilagus audubonii (Baird, 1858)
S. bachmani (Waterhouse, 1839)
Lepus californicus Gray, 1837

Order Rodentia
Family Sciuridae
Spermophilus (Otospermophilus) beecheyi (Richardson, 1823)

Family Geomyidae
Thomomys bottae (Eydoux and Gervais, 1836)

Family Heteromyidae
Dipodomys agilis Gambel, 1848
Perognathus californicus Merriam, 1889

Family Cricetidae
Reithrodontomys megalotus (Baird, 1858)
Peromyscus imperfectus Dice, 1925 †
Onychomys torridus (Coues, 1874)
Neotoma fuscipes Baird, 1858
Microtus californicus (Peale, 1848)

Order Carnivora
Family Mustelidae
Mustela frenata Lichtenstein, 1831
Taxidea taxus (Schreber, 1778)
Spilogale putorius (Linnaeus, 1758)
Mephitis mephitis (Schreber, 1776)

Family Canidae
Canis latrans Say, 1823
C. lupus Linnaeus, 1758
C. familiaris Linnaeus, 1758
C. dirus Leidy, 1858 †
Urocyon cinereoargenteus (Schreber, 1775)

Family Procyonidae
Procyon lotor (Linnaeus, 1758)
Bassariscus astutus (Lichtenstein, 1830)

Family Ursidae
Arctodus simus (Cope, 1879) †
Ursus americanus Pallas, 1780
U. arctos horribilis (Ord, 1815) ¶

Family Felidae
Smilodon fatalis (Leidy, 1868) †
S. fatalis brevipes (Merriam and Stock, 1932) †
Homotherium serum (Cope, 1893) †
Panthera atrox (Leidy, 1835) †
P. onca agusta (Simpson, 1941) †
Felis concolor Linnaeus, 1758 †
Felis sp. (small)
Lynx rufus (Shreber, 1777)

Order Proboscidea
Family Mammutidae
Mammut americanum (Kerr, 1791) †

Family Elephantidae
Mammuthus columbi (Falconer, 1857) †

Order Perissodactyla
Family Equidae
Equus cf. *E. occidentalis* Leidy, 1865 †
E. conversidens Owen, 1869 †

Family Tapiridae
Tapirus californicus Merriam, 1913 †

Order Artiodactyla
Family Tayassuidae
Platygonus cf. *P. compressus* LeConte, 1848 †

Family Camelidae
Camelops hesternus (Leidy, 1854) †
Hemiauchenia macrocephala (Cope, 1893) †

Family Cervidae
Odocoileus cf. *O. hemionus* (Rafinesque, 1817)
Cervus cf. *C. elaphus* (Linnaeus, 1758) ¶ ‡

Family Antilocapridae
Capromeryx minor Taylor, 1911 †
Antilocapra americana (Ord, 1815)

Family Bovidae
Euceratherium sp. cf. *E. collinum* Furlong and Sinclair, 1904 †
Bison latifrons (Harlan, 1825) †
B. antiquus Leidy, 1852 †
Ovis aries Linnaeus, 1758 ¶

Taxonomy and identifications follow Simpson, 1945, Stock, 1956, Kurtén and Anderson, 1980, and Savage and Russell, 1983.

BIBLIOGRAPHY

Abel, O. 1912. *Grundzüge der Paleaeobiologie der Wirbeltiere*. Stuttgart: E. Schweizerbart. 708 pp.

Abel, O. 1926. *Amerikafahrt. Eindrücke, boebachtungen und studien eines naturforschers auf eine reise nach Nordamerika und Westindien*. Jena: G. Fischer. 462 pp.

Agenbroad, L.D., and J.I. Mead. 1986. Large carnivores from Hot Springs Mammoth Site, South Dakota. *National Geographic Research* 2(4):508–516.

Akersten, W.A. 1979. Genesis of Rancho La Brea fossil deposits. *Abstracts, Annual Meeting Southern California Academy of Sciences* 56:28.

Akersten, W.A. 1980. Fossils in asphalt. *Science* 208:550.

Akersten, W.A. 1984. Metro Rail meets the Tar Pits. *Terra* 23(5):17–20.

Akersten, W.A. 1985. Canine function in *Smilodon* (Mammalia: Felidae: Machairodontinae). *Contributions in Science* 356:1–22.

Akersten, W.A. 1985. Of dragons and sabertooths. *Terra* 23(5):13–19.

Akersten, W.A. 1991. Entrapping pools or flowing streams at Rancho La Brea? Both! *Abstract, Annual Meeting California Academy of Sciences*, no. 1.

Akersten, W.A., T.M. Foppe, and G.T. Jefferson. 1984. Dietary samples from large extinct herbivores, Rancho La Brea. *Abstracts, Annual Meeting Southern California Academy of Sciences* 36:18.

Akersten, W.A., T.M. Foppe, and G.T. Jefferson. 1988. New source of dietary data for extinct herbivores. *Quaternary Research* 30(1):92–97.

Akersten, W.A., R.L. Reynolds, and A.E. Tejada-Flores. 1979. New mammalian records from the late Pleistocene of Rancho La Brea. *Bulletin of the Southern California Academy of Sciences* 78(29):141–143.

Akersten, W.A., C.A. Shaw, and G.T. Jefferson. 1983. Rancho La Brea: Status and future. *Paleobiology* 9(3):211–217.

Anderson, E. 1968. Fauna of the Little Box Elder Cave, Converse County, Wyoming. *University of Colorado Studies in Earth Sciences Series* 6:1–59.

Anonymous. 1871. A skull of a badger (*Taxidea*) from four feet beneath the surface at Los Angeles, California, in asphaltum beds. *Proceedings of the California Academy of Sciences* 4:139.

Anonymous. 1915. Rancho La Brea's fossil beds. *Standard Oil Bulletin* 2(11):3–6.

Anonymous. 1980. *Sabre-tooth cat: Official state fossil*. California Division of Mines and Geology, Note 38, 1 p.

Antonius, O. 1933. Über einen Pferdeschadel aus dem Rancho La Brea. *Verhandlinge Zoologische-Botanische Gesselenschaft, Wien* 83:39–40.

Arnold, R. 1907a. The Los Angeles oil district, southern California. *Bulletin of the United States Geological Survey*, no. 309:138–202.

Arnold, R. 1907b. Geology and oil resources of the Summerland District, Santa Barbara County California. *Bulletin of the United States Geological Survey*, no. 321:1–93.

Arnold, R., and R. Anderson. 1907. Metamorphism by combustion of the hydrocarbons in the oil-bearing shale of California. *Journal of Geology* 15(8):750–758.

Axelrod, D.I. 1967. Quaternary extinctions of large mammals. *University of California Publications in Geological Sciences* 74:1–42.

Bandy, O.L., and L. Marincovich, Jr. 1973. Rates of late Cenozoic uplift, Baldwin Hills, Los Angeles. *California Science* 181:653–654.

Barrows, A.G. 1970. New excavations for fossils at Rancho La Brea, Los Angeles. *Mineral Information Services* 23(1):14–15.

Bedrossian, T.L. 1975. Vertebrate fossils and the

history of animals with backbones. *California Geology* 28(11):234–259.

Berger, R., and W.F. Libby. 1966. UCLA radiocarbon dates V. *Radiocarbon* 8:467–497.

Berger, R., and W.F. Libby. 1968. UCLA radiocarbon dates VIII. *Radiocarbon* 10:402–416.

Berger, R., R. Protsch, R.L. Reynolds, C. Rozaire, and J. R. Sackett. 1971. New radiocarbon dates based on bone collagen of California paleoindians. *Contributions of the University Archaeology Research Facility*, Department of Anthropology, Berkeley 12:43–49.

Berta, A. 1985. The status of *Smilodon* in North and South America. *Contributions in Science* 370: 1–15.

Bischoff, J.L., and R.J. Rosenbauer. 1981. Uranium series dating of human skeletal remains from the Del Mar and Sunnyvale sites. *California Science* 213:1003–1005.

Blake, W.P. 1856. Report of the explorations in California for railroad routes to connect with the routes near the 35th and 32nd parallels of north latitude. *Reports of explorations and surveys to ascertain the most practical and economical route for a railroad from the Mississippi River to the Pacific Ocean*, Pt. 2, 5:76. Washington: B Tucker.

Bohlin, B. 1940. Food habit of the Machaerodonts, with special regard to *Smilodon*. *Bulletin of the Geological Institute of Upsala* 28:157–174.

Bohlin, B. 1947. The sabre-toothed tigers once more. *Bulletin of the Geological Institute of Upsala* 32:11–20.

Borell, A. E. 1936. A modern La Brea tar pit. *Auk* 53:298–300.

Bovard, J.F. 1907. Notes on Quaternary Felidae from California. *University of California Publications, Bulletin of the Department of Geological Sciences* 5(10):155–170.

Branning, T. 1977. Oily grave of extinction. *Union Oil Company, California Seventy Six* 56(5):32–36.

Brattstrom, B.H. 1953. The amphibians and reptiles from Rancho La Brea. *Transactions of the San Diego Society of Natural History* 11:365–392.

Brattstrom, B.H. 1954. The fossil pit-vipers (Reptilia: Crotalidae) of North America. *Transactions of the San Diego Society of Natural History* 3: 31–46.

Brattstrom, B.H. 1991. Paleoecology of Rancho La Brea: A clue to global warning and the drought? *Southern California Academy of Sciences, Abstracts for the Annual Meeting No. 23.*

Bright, M. 1965. California radiocarbon dates. *University of California Los Angeles Archaeological Survey* 7:363–375.

Bromage, T.G., and S. Shermis. 1981. The La Brea Woman (HC 1323): Descriptive analysis. *Society of California Archaeologists Occasional Papers* 3:59–75.

Brown, C.A. 1960. *Palynological techniques*. Baton Rouge: Louisiana State University, 188 pp.

Bryan, W.A. 1927. A Pleistocene park in the making. *Los Angeles Museum Graphic* 1(3):77–82.

Bryant, H.C. 1929. *Outdoor Heritage* Los Angeles: Powell Publishing Co. 465 pp.

Burroughs, H. 1938. Pleistocene reborn: Prehistoric beasts feature of Los Angeles park. *Nature* 31(6):329–332.

Camp, C.L. 1917. An extinct toad from Rancho La Brea. *University of California Publications, Bulletin of the Department of Geological Sciences* 10(17):287–292.

Camp, C.L. 1970. *Earth song, a prologue to history*. Palo Alto, California: American West Publishing Co., 192 pp.

Campbell, Jr., K.E., and E.P. Tonni. 1980. A new genus of teratorn from the Huayquerian of Argentina (Aves: Teratornithidae). In *Papers in avian paleontology honoring Hildegarde Howard*, ed. K.E. Campbell, Jr., 59–68 *Contributions in Science* 330.

Chandler, A.C. 1914. Antelopes in the fauna of Rancho La Brea. *Bulletin of the Geological Society of America* 25:155.

Chandler, A.C. 1916a. Notes on *Capromeryx* material from the Pleistocene of Rancho La Brea. *University of California Publications, Bulletin of the Department of Geological Sciences* 9(10):111–120.

Chandler, A.C. 1916b. A study of the skull and dentition of *Bison antiquus* Leidy, with special reference to material from the Pacific coast. *University of California Publications, Bulletin of the Department of Geological Sciences* 9(11):121–135.

Chaney, R.W. 1938. Paleoecological interpretations of Cenozoic plants in western North America. *Botanical Reviews* 4:359–391.

Chaney, R.W., and H.L. Mason. 1933. A Pleistocene flora from the asphalt deposits at Carpinteria, California. *Carnegie Institute of Washington Publications* 415, paper 111:45–49.

Churcher, C.S. 1984. The status of *Smilodontopsis* (Brown, 1908) and *Ischyrosmilus* (Merriam, 1918). A taxonomic review of two genera of sabretooth cats (Felidae, Machairodontinae). *Life Sciences Contributions Royal Ontario Museum* 140:1–59.

Clover, S.T. 1932. *A pioneer heritage*. Los Angeles: Saturday Night Publishing Co., 291 pp.

Compton, L.V. 1934. New bird records from the Pleistocene of Rancho La Brea. *Condor* 36(5): 221–222.

Compton, L.V. 1937. Shrews from the Pleistocene of the Rancho La Brea asphalt. *University of California Publications, Bulletin of the Department of Geological Sciences* 24(5):85–90.

Comstock, J.A. 1941. A glacial botanic garden. *Los Angeles County Museum Quarterly* 1(1):12–15.

Cox, S.M. 1979. Preliminary census of the Ursidae from Rancho La Brea. *Abstracts, Annual Meeting Southern California Academy of Sciences* 89:45.

Cox, S.M. 1991. Size range or sexual dimorphism in *Arctodus simus* from Rancho La Brea. *Abstract, Annual Meeting California Academy of Sciences No. 15.*

Cox, S.M., and G.T. Jefferson. 1988. The first individual skeleton of *Smilodon* from Rancho La Brea. *Current Research in the Pleistocene* 5:66–67.

Crowder, R.E., and R.A. Johnson. 1963. Recent developments in the Jade-Buttram area of the Salt Lake Oil Field. *California Oil and Gas, Summary of Operations* 49(1):53–60.

Das, S., A. Doberenz, and R.W.G. Wyckoff. 1967. The lipids in fossils. *Comparative Biochemistry and Physiology* 23:519–525.

Davidson, A. 1914. The oldest known tree. *Bulletin of the Southern California Academy of Sciences* 13(1):14–16.

Dawson, W.R. 1948. Records of fringillids from the Pleistocene of Rancho La Brea. *Condor* 50(2):57–63.

Deevey, E.S., L.J. Gralenski, and V. Hoffren. 1959. Yale natural radiocarbon measurements, IV. *Radiocarbon* 1:144–172.

DeMay, I.S. 1941a. Quaternary bird life of the McKittrick asphalt, California. *Carnegie Institute of Washington Publications* 530:35–60.

DeMay, I.S. 1941b. Pleistocene bird life of the Carpinteria asphalt, California. *Carnegie Institute of Washington Publications* 530:61–76.

De Mofras, D. 1844. *Exploration du territoire de l'Oregon, des Californies et de la Mer Vermeille executée pendant les années 1840, 1841, et 1842*. Paris: A. Bertrand.

Denton, W. 1875. On the asphalt bed near Los Angeles, California. *Proceedings of the Boston Society of Natural History* 18:185–186.

Diamond, J.M. 1986. How great white sharks, sabretooth cats and soldiers kill; animal behavior. *Nature* 322:773–774.

Dice, L.R. 1925. Rodents and lagomorphs of the Rancho La Brea deposits. *Carnegie Institute of Washington Publications* 349:119–130.

Doberenz, A.R. 1967. Ultrastructure of fossil dentinal collagen. *California Tissue Research* 1:166–169.

Doberenz, A.R., and P. Matter III. 1965. Nitrogen analysis of fossil bones. *Comparative Biochemistry and Physiology* 16:253–258.

Doberenz, A.R., and R.W.G. Wyckoff. 1967. Fine structure in fossil collagen. *Proceedings of the National Academy of Sciences* 57:539–541.

Dorr, J.A., Jr. 1956. Terrors of the tar pits—two toenail sketches. *Michigan Alumni Quarterly Review* 62(18):220–224.

Douglas, D. 1952. Measuring low-level radioactivity. *General Electric Review* 55:16–20.

Doyen, J.T., and S.E. Miller. 1980. Review of Pleistocene darkling ground beetles of the California asphalt deposits (Coleoptera: Tenebrionidae: Zopheridae). *Pan-Pacific Entomology* 56(1):1–10.

Duque, J., and L.G. Barnes. 1975. *Smilodon*, is this how you looked? *Terra* 14(1):18–24.

Eakle, A.S. 1923. Minerals of California. *California State Mines Bureau Bulletin* 91:236.

Eaton, J.E. 1928. Divisions and duration of the Pleistocene in southern California. *Bulletin of the American Association of Petroleum Geologists* 12(2):111–141.

Eberhart, H. 1961. The cogged stones of southern California. *American Antiquity* 26(3):361–370.

Edholm, C.L. 1914. Prehistoric man of Los Angeles. *Technical World Magazine* 22(2):216–217.

Emerson, S.B., and L. Radinsky. 1980. Functional analysis of sabertooth cranial morphology. *Paleobiology* 6(3):295–312.

Emslie, S.D. 1988. The fossil history and phylogenetic relationships of condors (Ciconiiformes: Vulturidae) in the New World. *Journal of Vertebrate Paleontology* 8(2):212–228

Emslie, S.D., and N.J. Czaplewski. 1985. A new record of giant short-faced bear, *Arctodus simus*, from western North America with a re-evaluation of its paleobiology. *Contributions in Science* 371:1–12.

Engles, W.L. 1935. Status of *Toxostoma redivivum* in the Rancho La Brea fauna. *Condor* 37:258.

Engstrand, I.H.W. 1989. California's Ranchos: An enduring legacy. *Terra* 28(2):6–15.

Esterly, C.O. 1913. The "oil fly" of Southern California. *Bulletin of the Southern California Academy of Sciences* 12(1):9–11.

Fay, L.P. 1991. First record of tortoise (Reptilia: Testudinae) from Rancho La Brea. *Southern California Academy of Sciences, Abstracts for the Annual Meeting* No. 8.

Fergusson, G.J., and W.F. Libby. 1963. UCLA radiocarbon dates II. *Radiocarbon* 5:1–22.

Fisher, H.I. 1944. The skulls of the cathartid vultures. *Condor* 46:272–296.

Fisher, H.I. 1945. Locomotion in the fossil vulture *Teratornis*. *The American Midland Naturalist* 33(3):725–742.

Fisher, H.I. 1947. The skeletons of Recent and fossil *Gymnogyps*. *Pacific Science* 1(4):227–236.

Frick, C. 1937. The horned ruminants of North America. *Bulletin of the American Museum of Natural History* 69:1–669

Frost, F.H. 1927. The Pleistocene flora of Rancho La Brea. *University of California Publications in Botany* 14(3):73–98.

Frost, R. 1925. The pits of Rancho La Brea. *Mentor* 13(7):19–21.

Furlong, E.L. 1931. Distribution and description of skull remains of the Pliocene antelope *Sphenophalos* from the northern Great Basin Province. *Carnegie Institute of Washington Publications* 418:27–36.

Furlong, E.L. 1946. Generic identification of the Pleistocene antelope from Rancho La Brea. *Carnegie Institue of Washington Publications* 551:137–140.

Gagne, R.J., and S.E. Miller. 1982. *Protochrysomyia howardi* from Rancho La Brea, California, Pleistocene, new junior synonym of *Cochliomyia macellaria* (Diptera: Calliphoridae). *Bulletin of the Southern California Academy of Sciences* 80(2):95–96.

Gasteiger, L.D. 1972. Skullduggery in La Brea. *Cosmos* 6(8):20–22.

Gilbert, J.Z. 1909. Finds in the "fossil gardens" of Los Angeles. *Los Angeles Museum Graphic*, 30 Oct., 1–4.

Gilbert, J.Z. 1910. The fossils of Rancho La Brea. *Bulletin of the Southern California Academy of Sciences* 9(1):11–51.

Gilbert, J.Z. 1927. The bone drift in the tar beds of Rancho La Brea. *Bulletin of the Southern California Academy of Sciences* 26(3):59–66.

Giles, E. 1960. Multivariate analysis of Pleistocene and Recent coyotes (*Canis latrans*) from California. *University of California Publications in Geological Sciences* 36(8):369–390.

Ginnett, T.F., and C.L. Douglas. 1982. Food habits of feral burros and desert bighorn sheep in Death Valley National Monument. *Desert Bighorn Council, Transactions 1982*, 81–86.

Ginsburg, L. 1961. Plantigradie et digitigradie chez les Carnivores Fissipèdes. *Mammalia* 25(1):1–21.

Goldin, J. 1979. *Bison antiquus* at Rancho La Brea. *Abstracts, Annual Meeting Southern California Academy of Sciences* 90:45.

Goldman, E.A. 1944. Classification of wolves. Part II. In *The wolves of North America*, ed. S.P. Young and E.A. Goldman, 389–636. Washington, D.C.: American Wildlife Institute.

Goldman, E.A. 1946. Classification of the races of the puma. Part II. In *The puma: Mysterious American cat*, ed. S.P. Young and E.A. Goldman, 177–302. Washington, D.C.: American Wildlife Institute.

Gonyea, W.J. 1976a. Adaptive differences in the body proportions of large felids. *Acta Anatomica* 96:81–96.

Gonyea, W.J. 1976b. Behavioral implications of saber-toothed felid morphology. *Paleobiology* 2(4): 332–342.

Graham, R.W. 1976b. Pleistocene and Holocene mammals, taphonomy and paleoecology of the Freisenhahn Cave local fauna, Bexar County, Texas. Ph.D. diss., University of Texas, Austin.

Graham, R.W., and E.L. Lundelius. 1984. Coevolutionary disequilibrium and Pleistocene extinctions. In *Quaternary extinctions: A prehistoric revolution*, ed. P.S. Martin and R.G. Klein, 223–249. Tucson: University of Arizona Press.

Grant, U.S., and W.E. Sheppard. 1939. Some recent changes of elevation in the Los Angeles basin of southern California, and their possible significance. *Bulletin of the Seismological Society of America* 29(2):299–326.

Grayson, D.K. 1977. Pleistocene avifaunas and the overkill hypothesis. *Science* 195:691–693.

Grinnell, J. 1933. Review of the Recent mammal fauna of California. *University of California Publications in Zoology* 40:71–231.

Grinnell, F., Jr., 1908. Quaternary myriopods and insects of California. *University California Publications, Bulletin of the Department of Geological Sciences* 5(12):207–215.

Guilday, J.E. 1984. Pleistocene extinction and environmental change: Case study of the Appalachians. In *Quaternary extinctions: A prehistoric revolution*, ed. P.S. Martin and R.G. Klein, 250–258. Tucson: University of Arizona Press.

Gust, S. 1991. Age and sex distribution of the Rancho La Brea horse based on pelvic characters. *Abstracts, Journal of Vertebrate Paleontology*, 11(suppl. to no. 3):33A.

Gust, S., and H. Howard. 1991. Change in bird species through time at Rancho La Brea: Ecological or taphonomic causes? *Abstract, Annual Meeting California Academy of Sciences* No. 9.

Gust, S., and E. Scott. 1989. Morphological indicators of paleoenvironment: The Rancho La Brea horse through time. *Abstracts, Journal of Vertebrate Paleontology* (suppl. to no. 3) 9(3):24A.

Guthrie, R.D. 1984. Mosaics, allelochemics and nutrients: An ecological theory of late Pleistocene megafaunal extinctions. In *Quaternary extinctions: A prehistoric revolution*, ed. P.S. Martin and R.G. Klein, 259–298. Tucson: University of Arizona Press.

Gutman, T.E. 1979. The use of asphaltum sourcing in archaeology. *Journal of New World Archaeology* 3(2):32–43.

Hall, E.R. 1936. Mustelid mammals from the Pleistocene of North America with systematic notes of some recent members of the genera *Mustela*, *Taxidea*, and *Mephitis*. *Carnegie Institute of Washington Publications* 473:41–119.

Hallett, M. 1987. Two wolves, one land: Could two closely related social predators share the same environment? *Terra* 25(6):11–17.

Hancock, G.A. 1964. *A pictorial account of one man's score in fourscore years*. San Jose, California: Paramount Printing Co., 281 pp.

Hanks, H.G. 1882. *On the occurrence of vivianite in Los Angeles County*. State Mineralogist of California, Second Annual Report, part 1, p. 265.

Hanna, G.D. 1924. *Succinea avara* Say, from the tar pits of California. *Nautilus* 37(3):106.

Hanna, G.D. 1949. Animals and oil traps. *Wasmann Collector* 7(4):138.

Hansen, R.M. 1976. Foods of free-roaming horses in southern New Mexico. *Journal of Range Management* 29:347.

Harris, J.M. 1989. Trapped in time: Restoration of Charles R. Knight mural. *Terra* 28(2):36–37.

Harris, J.M., and G.T. Jefferson. 1985. *Rancho La Brea: Treasures of the tar pits*. Science Series, no. 31. Los Angeles: Natural History Museum of Los Angeles County, 87 pp.

Hay, O.P. 1927. The Pleistocene of the western region of North America and its vertebrated animals. *Carnegie Institute of Washington Publications* 322B:1–346.

Haynes, G., and D. Stanford. 1984. On the possible utilization of *Camelops* by early man in North America. *Quaternary Research* 22:216–230.

Heald, F.P. 1986. Paleopathology at Rancho La Brea. *Anthroquest* 36:6–7.

Heald, F.P. 1989. Injuries and diseases in *Smilodon californicus* Bovard, 1904, (Mammalia, Felidae) from Rancho La Brea, California. *Abstracts, Journal of Vertebrate Paleontology* (suppl. to no. 3) 9(3):24A.

Heald, F.P., and C.A. Shaw. 1991. Sabertooth cats. In *Great cats: Majestic creatures of the wild*, ed. Seidensticker, J and S. Lumpkin, 26–27. Emmaus, Pennsylvania: Rodale Press, Inc.

Hebert, G.J. 1989. Rancho La Brea today. *The Chesopiean, Chesopiean Library of Archaeology* 271:21–26.

Heizer, R.F. 1943. Aboriginal use of bitumen by the California Indians. *Bulletin of the California Division of Mines* 118:74.

Heizer, R.F. 1953. Sites attributed to early man in California. *University of California Los Angeles, Archaeology Surveys* 21:1–4.

Heizer, R.F., and A.E. Treganza. 1944. Mines and quarries of the Indians of California. *California Division of Mines* 40(3):333.

Heric, T.M. 1969. Rancho La Brea: Its history and its fossils. *Journal of the West* 8(2):209–230.

Hibbard, C.W. 1958. Summary of North America Pleistocene mammalian local faunas. *Michigan Academy of Sciences, Arts and Letters* 43:1–32.

Hill, D. 1991. Was dire wolf a bone crusher? A morphological comparison of *Canis dirus* and *Crocuta crocuta*. *Abstract, Annual Meeting California Academy of Sciences* No. 13.

Ho, T.Y. 1965. The amino acid composition of bone and tooth proteins in late Pleistocene mammals. *Proceedings of the National Academy of Sciences* 54(1):26–31.

Ho, T.Y. 1966. The isolation and amino acid composition of the bone collagen in Pleistocene mammals. *Comparative Biochemistry and Physiology* 18:353–358.

Ho, T.Y. 1967. Relationship between amino acid contents of mammalian bone collagen and body temperature as a basis for estimation of body temperature of prehistoric mammals. *Comparative Biochemistry and Physiology* 22:113–119.

Ho, T.Y. 1967. The amino acids of bone and dentine collagens in Pleistocene mammals. *Biochimica Biophysica Acta* 133:568–573.

Ho, T.Y., Marcus, L.F., and R. Berger. 1969. Radiocarbon dating of petroleum-impregnated bone from tar pits at Rancho La Brea. *California Science* 164:1051–1052.

Hodgson, S.F. 1980. Onshore oil and gas seeps in California. *California Division of Oil and Gas* TR26:1–97.

Hoffstetter, R. 1950. Algunas observaciones sobre los caballos fosiles de la America del Sur. *Amerhippus* gen. nov. *Boletin Informaciones Cientificas Nacionales* 3:426–454.

Hoots, H.W. 1931. *Geology of the eastern part of the Santa Monica Mountains, Los Angeles County, California*. U.S. Geological Survey, Professional Papers 165C:1–134.

Hornaday, W.T. 1924. *Tales from nature's wonderlands*. New York: C. Scribner. 235pp.

Howard, H. 1923. The fossil beds of La Brea. *International Association of High School Natural History Clubs*, 2 Nov., 1.

Howard, H. 1927. A review of the fossil bird, *Parapavo californicus* (Miller), from the Pleistocene asphalt beds of Rancho La Brea. With an appendix; statistical identification as applied to *Parapavo*, by F.H. Frost. *University of California Publications, Bulletin of the Department of Geological Sciences* 17(1):1–62.

Howard, H. 1928. The beak of *Parapavo californicus* (Miller). *Bulletin of the Southern California Academy of Sciences* 27(3):90–91.

Howard, H. 1929. Additional bird records from the Pleistocene of Rancho La Brea. *Condor* 31: 251–252.

Howard, H. 1930. A census of the Pleistocene birds of Rancho La Brea from the collection of the Los Angeles museum. *Condor* 32:81–88.

Howard, H. 1932. Eagles and eagle-like vultures of the Pleistocene of Rancho La Brea. *Carnegie Institute of Washington Publications* 429:1–82.

Howard, H. 1933. A new species of owl from the Pleistocene of Rancho La Brea, California. *Condor* 35:66–69.

Howard, H. 1935. The Rancho La Brea wood ibis. *Condor* 37(5):251–253.

Howard, H. 1936. Further studies upon the birds of the Pleistocene of Rancho La Brea. *Condor* 38(1):32–36.

Howard, H. 1937. A Pleistocene record of the passenger pigeon in California. *Condor* 39(1):12–14.

Howard, H. 1938. The Rancho La Brea caracara: A new species. *Carnegie Institute of Washington Publications* 487(5):217–240.

Howard, H. 1939. Aves. *Fortschritte der Palaeontologie* 2 (1937–1938):309–322.

Howard, H. 1940. A new race of caracara from the Pleistocene of Mexico. *Condor* 42:41–44.

Howard, H. 1941. A review of the American fossil storks. *Carnegie Institute of Washington Publications* 530(7):187–203.

Howard, H. 1945. *Fossil birds*. Science Series no. 10, Paleontology no. 6. Los Angeles: Los Angeles County Museum, 40 pp.

Howard, H. 1945. Observations on young tarsometatarsi of the fossil turkey *Parapavo californicus* (Miller). *Auk* 62:596–603.

Howard, H. 1946. A review of the Pleistocene birds of Fossil Lake, Oregon. *Carnegie Institute of Washington Publications* 551:141–195.

Howard, H. 1947a. A preliminary survey of trends in avian evolution from Pleistocene to Recent time. *Condor* 49(1):10–13.

Howard, H. 1947b. An ancestral golden eagle raises a question in taxonomy. *Auk* 64:287–291.

Howard, H. 1949. The California turkey. *Los Angeles County Museum Quarterly* 7(4):15–16.

Howard, H. 1950. *Teratornis*: The wonder bird of the Ice Age. *Los Angeles County Museum Leaflet Series in Science*, no. 3:1–4.

Howard, H. 1952. The prehistoric avifauna of Smith Creek Cave, Nevada, with a description of a new gigantic raptor. *Bulletin of the Southern California Academy of Sciences* 51(2):50–54.

Howard, H. 1953. Forty years at Rancho La Brea. *Los Angeles County Museum Quarterly* 10(2):6–12.

Howard, H. 1955. *Fossil birds*. Science Series no. 17, Paleontology no. 10. Los Angeles: Los Angeles County Museum, 40 pp.

Howard, H. 1960. Significance of carbon-14 dates for Rancho La Brea. *Science* 131:712–714.

Howard, H. 1961. *Fossil birds*. Science Series, no. 17, Paleontology no. 10. Los Angeles: Los Angeles County Museum. 44 pp.

Howard, H. 1962. A comparison of prehistoric avian assemblages from individual pits at Rancho La Brea, California. *Contributions in Science* 58:1–24.

Howard, H. 1964. A fossil owl from Santa Rosa Island, California, with comments on the eared owls of Rancho La Brea. *Bulletin of the Southern California Academy of Sciences* 63(1):27–31.

Howard, H. 1964. Fossil Anseriformes. In *The waterfowl of the world*, ed. J. Delacour, 4:233–326. London: Country Life Limited .

Howard, H. 1968. Limb measurements of the extinct vulture, *Coragyps occidentalis*. *Papers of the Archaeological Society of New Mexico* 1:115–128.

Howard, H. 1972. The incredible Teratorn again. *Condor* 74(3):341–344.

Howard, H. 1972. Type specimens in the collections of the Natural History Museum of Los Angeles County. *Contributions in Science* 228:1–27.

Howard, H. 1974. Postcranial elements of the extinct Condor *Breagyps clarki* (Miller). *Contributions in Science* 256:1–24.

Howard, H., and A.H. Miller. 1939. The avifauna associated with the human remains at Rancho La Brea, California. *Carnegie Institute of Washington Publications* 514:39–48.

Hrdlicka, A. 1918. Recent discoveries attributed to early man in America. *Smithsonian Instution Bureau of American Ethnology Bulletin* 66:9–67.

Hubbs, C.L., G.S. Bien, and H.E. Seuss. 1960. La Jolla natural radiocarbon measurements, I. *Radiocarbon* 2:197–223.

Hubbs, C.L., and B.W. Walker. 1947. Abundance of desert animals indicated by capture in fresh road tar. *Ecology* 28(4):464–466.

Husband, R.A. 1924. Variability in *Bubo virginianus* from Rancho La Brea. *Condor* 26:220–225.

Jackson, H.H.T. 1951. Classification of the races of coyote. Part II. In *The clever coyote*, ed. S.P. Young and H.H.T. Young, 227–441. Washington, D.C.: American Wildlife Institute.

Jaffe, E.B., and A.M. Sherwood. 1951. Physical and chemical comparison of modern and fossil tooth and bone material. *U.S. Geological Survey Reports* TEM-149:1–19.

Jefferson, G.T. 1983. First record of jaguar from the late Pleistocene of California. *Bulletin of the Southern California Academy of Sciences* 82(2):95–98.

Jefferson, G.T. 1986. A new specimen of tapir from Rancho La Brea. *Current Research in the Pleistocene* 3:68–69.

Jefferson, G.T. 1986. Late Pleistocene large mammalian herbivores: Implications for big game hunters in southern California. *Abstracts, Southern California Academy of Sciences* 5.

Jefferson, G.T. 1987. Fossil tapirs from the late Cenozoic of western North America. *Abstracts, Southern California Academy of Sciences* 13.

Jefferson, G.T. 1988. Late Pleistocene large mammalian herbivores: Implications for big game hunters in southern California. In *Desert ecology 1986: A research symposium*, ed. R.G. Zahary, 15–35. Southern California Academy of Sciences and Southern California Desert Studies Consortium; Los Angeles.

Jefferson, G.T. 1989. Digitized sonic location and computer imaging of Rancho La Brea specimens from the Page Museum salvage. *Current Research in the Pleistocene* 6:45–47.

Jefferson, G.T. 1989. Late Cenozoic tapirs (Mammalia: Perissodactyla) of western North America. *Contributions in Science* 406:1–22.

Jefferson, G.T. 1989. Taphonomic measurements with a sonic digitizer at Rancho La Brea. *Abstracts, Journal of Vertebrate Paleontology* (suppl. to no. 3) 9(3):27A.

Jefferson, G.T. 1991. Size and sexual dimorphism in *Panthera leo atrox* (Mammalia; Felidae) from Rancho La Brea. *Abstract, Annual Meeting California Academy of Sciences* no. 16.

Jefferson, G.T., and S.M. Cox. 1986. New articulated vertebrate remains from Rancho La Brea. *Current Research in the Pleistocene* 3:70–71.

Jefferson, G.T., and J. Goldin. 1989. Seasonal migration of *Bison antiquus* from Rancho La Brea. *Quaternary Research* 31:107–112.

Jerison, H.J. 1973. *Evolution of the brain and intelligence*. New York: Academic Press, 482 pp.

Johnson, D.L. 1977a. The California Ice-Age refugium and the Rancholabrean extinction problem. *Quaternary Research* 8:149–153.

Johnson, D.L. 1977b. The late Quaternary climate of coastal California: Evidence for an Ice Age refugium. *Quaternary Research* 8:154–179.

Jordan, D.S., and H. Hannibal. 1923. Fossil sharks and rays of the Pacific slope of North America. *Bulletin of the Southern California Academy of Sciences* 22(2):27–63.

Keller, J.S., and D.F. McCarthy. 1989. Data recovery at the Cole Canyon site (CA-RIV-1139), Riverside County, California. *Pacific Coast Archaeological Society Quarterly* 25(1):1–90.

Kellogg, L. 1912. Pleistocene rodents of California. *University of California Publications, Bulletin of the Department of Geology* 7(8):151–168.

Kelly, W.A. 1940. Tar as a trap for unwary humans. *American Journal of Science* 238:451–452.

Kennedy, G.E. 1989. A note on the ontogenetic age of the Rancho La Brea hominid, Los Angeles, California. *Bulletin of the Southern California Academy of Sciences* 88(3):123–126.

King, G. 1953. The mystery of the La Brea tar pits. *Sir Magazine* 10(7):1–4.

Klide, A.M. 1989. Overriding vertebral spinous processes in the extinct horse, *Equus occidentalis*. *American Journal of Veterinary Research* 50(4): 592–593.

Knowlton, F.H. 1916. Notes on two conifers from the Pleistocene Rancho La Brea asphalt deposits near Los Angeles, California. *Journal of the Washington Academy of Sciences* 6:85–86.

Kraglievich, L. 1926. Los arcterios Norteamericanos en relacion con los de Sud America. *Anales del Museo Nacional de Historia Natural, Buenos Aires* 34:1–16.

Kraglievich, L. 1928. *Mylodon darwini* Owen es la especiegenotipo de Mylodon Owen, rectificacion de la nomenclatura generica de los milodontes. *Revista de la Sociedad Argentina de Ciencias Naturales* 9:169–185.

Kroeber, A.L. 1962. The Rancho La Brea skull *American Antiquity* 27(3):416–417.

Kunz, G.F. 1916. *Ivory and the elephant in art, in archeology, and in science.* Garden City, N.Y.: Doubleday, Page and Co., 527 pp.

Kurtén, B. 1960. A skull of the grizzly bear (*Ursus arctos* L.) from Pit 10, Rancho La Brea. *Contributions in Science* 39:1–7.

Kurtén, B. 1984. Geographic differentiation in the Rancholabrean dire wolf (*Canis dirus* Leidy) in North America. In *Contributions in Quaternary vertebrate paleontology: A volume in memorial to John E. Guilday*, ed. H. H. Genoways and M. R. Dawson, Carnegie Museum of Natural History Special Publication, 8:218–227.

Kurtén, B. 1986. Tar Pit Fossils. *Boreas Book Reviews, Boreas* 15:82.

Kurtén, B., and E. Anderson. 1980. *Pleistocene mammals of North America.* New York: Columbia University Press, 443 pp.

Kvenvolden, K.A., and E. Peterson. 1972. Amino acids in late Pleistocene bone from Rancho La Brea, California. *Abstracts, Geological Society of America* 5(7):704–705.

Lackie, M.O. 1949. Ancient secrets of Rancho La Brea. *Union Oil California On Tour* 11(3):12–17.

LaDuke, T.C. 1983. The fossil snake fauna of Pit 91, Rancho La Brea, Los Angeles County, California. Master's thesis, Michigan State University, 57 pp.

LaDuke, T.C. 1991. The fossil snakes of Pit 91, Rancho La Brea, California. *Contributions in Science* 424:1–28.

LaDuke, T.C. 1991. First record of salamander remains from Rancho La Brea. *Abstract, Annual Meeting California Academy of Sciences* no. 7.

Lamb, R.V. 1988. Nonmarine mollusks of Pit 91, Rancho La Brea, and their paleoecologic implications. *Abstracts Annual Meeting Southern California Academy of Sciences*, no. 119.

Lamb, R.V. 1989. The nonmarine mollusks of Pit 91, Rancho La Brea, southern California, and their paleoecologic and biogeographic implications. Master's thesis, California State University, Northridge, 365 pp.

Lamb, R.V. 1991. The molluscan paleontology and paleoecology of Pit 91, Rancho La Brea, Southern California. *Abstract, Annual Meeting California Academy of Sciences* No. 5.

Lamb, R.V. and G.T. Jefferson. 1988. Mollusks associated with articulated vertebrate remains from Rancho La Brea. *Current Research in the Pleistocene* 5:59–60.

Lamb, R.V. and N.J. Maloney. 1991. The depositional environments, history and correlation of the sediments of Pit 91, Rancho La Brea, southern California. *Southern California Academy of Sciences, Abstracts for the Annual Meeting* no. 2.

Larew, H. 1987. Two cynipid wasp acorn galls preserved in the La Brea tar pits (early Holocene). *Proceedings of the Entomological Society of Washington* 89(4):831–833.

Larson, L.M. 1930. Osteology of the California roadrunner Recent and Pleistocene. *University of California Publications in Zoology* 32(4):409–428.

Laudermill, J.D., and P.A. Munz. 1934. Plants in the dung of *Nothrotheriops* from Gypsum Cave, Nevada. *Carnegie Institute of Washington Publications* 453(4):29–37.

Laudermill, J.D., and P.A. Munz. 1938. Plants in the dung of *Nothrotheriops* from Rampart and Mauv Caves, Arizona. *Carnegie Institute of Washington Publications* 487(7):271–281.

LeConte, J. 1882. On certain remarkable tracks found in the rocks of Carson Quarry. *Proceedings of the California Academy of Science*, 27 August, 1–10.

Lepper, B.T., T.D. Frolking, D.C. Fisher, G. Goldstein, J.E. Sanger, D.A. Wymer, J.G. Ogden, and P.E. Hooge. 1991. Intestinal contents of a late Pleistocene mastoidont from midcontinental North America. *Quaternary Research* 36:120–125.

Lonnberg, E. 1921. Some speculations on the origin of the North American ornithic fauna. *Kungliga Svenska Vetenskapsakadamiens Handlingar* (3)4, no. 6:1–24.

Lummis, C.F. 1925. *Mesa, cañon and pueblo; our wonderland of the Southwest, its marvels of nature, its pageant of the earth building, its strange peoples, its centuried romance.* New York: The Century Co., 517 pp.

Lundelius, E.L. 1960. *Mylohyus nastus*, long-nosed peccary of the Texas Pleistocene. *Bulletin of the Texas Memorial Museum* 1:1–40.

Lundelius, E.L., T. Downs, E.H. Lindsay, H.A. Semken, R.J. Zakrewski, C.S. Churcher, C.R. Harington, G.E. Schultz, and S.D. Webb. 1987. The North American Quaternary sequence. In *Cenozoic mammals of North America*, ed., M.O. Woodburne, 211–235. Berkeley: University of California Press.

Lyon, G.M. 1938. *Megalonyx milleri*, a new Pleistocene ground sloth from southern California. *Transactions of the San Diego Society of Natural History* 9(6):15–30.

Lytle, J.W. 1926. The Rancho La Brea asphalt pits. *Los Angeles Museum, Museum Graphic* 1(1):23.

Macdonald, J.R. 1969. The deadly asphalt pits. *National Wildlife Magazine* August/September: 25–28.

Macdonald, J.R., and L.J. Macdonald. 1974. *Sabretooth cats and imperial mammoths: A guidebook to fossil hunting in southern California.* Los Angeles: The Ward Richie Press, 112 pp.

Macdonald, J.R., and G. Sibley. 1969. Paleopathological ponderings or how to tell a sick sabertooth. *Los Angeles County Museum Quarterly* 8(2):26–30.

MacPherson, A.H. 1965. The origin of diversity in mammals of the Canadian Arctic tundra. *Systematic Zoology* 14:153–173.

Maloney, N.J. 1970. Late Pleistocene sedimentation and asphalt entrapment, Rancho La Brea. *Abstracts, Annual Meeting Southern California Academy of Sciences* 5–6 May, 28.

Maloney, N.J. 1979. Late Pleistocene sedimentation and asphalt emplacement, Rancho La Brea. *Abstracts, Annual Meeting Southern California Academy of Sciences* 55:28.

Maloney, N.J., and W.A. Akersten. 1973. Field trip to La Brea tar pits and L.A. County Museum of Natural History, field trip #4. *Association of Engineering Geologists, 1973 Annual Meeting. Field Trips Guidebook*, 1–7.

Maloney, N.J., and W.A. Akersten. 1976. Formation of calcareous sandstone at asphalt-groundwater contacts in fluvial sediments, Rancho La Brea, California. *Abstracts, Geological Society of America Cordilleran Section* 8(3):393.

Maloney, N.J., J.J. Criscione, and L.L. Bramlett. 1973. Fluvial sedimentation at Rancho La Brea. *Abstracts, Geological Society of America Cordilleran Section* 5(1):77.

Maloney, N.J., and J. Daigh. 1971. Sediment facies in Pit 91, Rancho La Brea tar pits, California. *Abstracts, Geological Society of America Cordilleran Section* 3(2):156–157.

Maloney, N.J., J.K. Warter, and W.A. Akersten. 1974. Probable origin of the fossil deposits in Pit 91, Rancho La Brea tar pits, California. *Abstracts, Geological Society of America Cordilleran Section* 6(3):212.

Marcus, L.F. 1960. A census of the abundant large Pleistocene mammals from Rancho La Brea. *Contributions in Science* 38:1–11.

Marcus, L.F. 1979. Radiocarbon dates for Rancho La Brea. *Abstracts, Annual Meeting Southern California Academy of Sciences* 57:29.

Marcus, L.F. 1991. Rancho La Brea research—a retrospective and prospective view. *Abstract, Annual Meeting California Academy of Sciences* no. 24.

Marcus, L.F., and R. Berger. 1984. The significance of radiocarbon dates for Rancho La Brea. In *Quaternary extinctions: A prehistoric revolution,* ed. P.S. Martin and R.G. Klein, 159–183. Tucson: University of Arizona Press.

Marsh, O.C. 1883. On the supposed human footprints recently found in Nevada. *American Journal of Science* 26:139–140.

Martin, P.S. 1967. Prehistoric overkill. In *Pleistocene extinctions: The search for a cause,* ed. P.S. Martin and H.E. Wright, Jr., 75–120. New Haven: Yale University Press.

Martin, P.S. 1984. Prehistoric overkill: The global model. In *Quaternary extinctions: A prehistoric revolution,* ed. P.S. Martin and R.G. Klein, 354–403. Tucson: University of Arizona Press.

Martin, R.A. 1974. Fossil mammals from the Colman IIA fauna, Sumter County. In *Pleistocene mammals of Florida,* ed. S.E. Webb, 35–99. Gainesville: University of Florida Press.

Mason, H.L. 1927. Fossil records of some west American conifers. *Carnegie Institute of Washington Publications* 346(5):139–160.

Matthew, W.D. 1913. The asphalt group of fossil skeletons. The tar pits of Rancho La Brea, California. *American Museum Journal* 13:291–297.

Matthew, W.D. 1916a. The grim wolf of the tar pits. The great extinct wolf from the asphalt deposits at Rancho La Brea near Los Angeles. Skeleton of *Canis dirus* recently mounted in the American Museum. *American Museum Journal* 16(1):45–47.

Matthew, W.D. 1916b. Scourge of the Santa Monica Mountains. *American Museum Journal* 16(7): 468–472.

McDonald, H.G. 1991. Sexual dimorphism in the skull of Harlan's ground sloth. *Abstract, Annual Meeting California Academy of Sciences* no. 10.

McDonald, J.N. 1979. New evidence of *Bison latifrons* in the Rancho La Brea fauna (Los Angeles County, California). *Abstract, Journal of Colorado-Wyoming Academy of Sciences* 11(196): 88.

McDonald, J.N. 1981. *North American bison, their classification and evolution.* Berkeley: University of California Press, 316 pp.

McMenamin, M.A.S., D.J. Blunt, K.A. Kvenvolden, S.E. Miller, L.F. Marcus, and R. Pardi. 1982. Amino acid geochemistry of fossil bones from the Rancho La Brea asphalt deposit, California. *Quaternary Research* 18:174–183.

Meier, D.K., and K.J. Rheinhard. 1990. Preliminary observation of pathological conditions of dire wolves of the Page Museum. *Journal of Vertebrate Paleontology,* supplement to 10(3):35A.

Menard, H.W., Jr. 1947. Analysis of measurements in length of the metapodials of *Smilodon. Bulletin of the Southern California Academy of Sciences* 46(3):127–135.

Merriam, J.C. 1906. Recent discoveries of Quaternary mammals in southern California. *Science* 24:248–250.

Merriam, J.C. 1908. Death trap of the ages. *Sunset Magazine* 21(6):465–475.

Merriam, J.C. 1908. Fauna of the asphalt beds exposed near Los Angeles, California. *Abstracts, Bulletin of the Geological Society of America* 18: 659.

Merriam, J.C. 1909a. The skull and dentition of

an extinct cat closely allied to *Felis atrox* Leidy. *University of California Publications, Bulletin of the Department of Geology* 5(20):291–304.

Merriam, J.C. 1909b. A death-trap which antedates Adam and Eve: The discovery of a Californian tar-swamp that holds the bones of extinct monsters. *Harper's Weekly*, no. 53 (Dec.), 11–12.

Merriam, J.C. 1910. New mammalia from Rancho La Brea. *University of California Publications, Bulletin of the Department of Geology* 5(25):391–395.

Merriam, J.C. 1911a. The fauna of Rancho La Brea, Pt. 1, occurrence. *Memoirs of the University of California* 1(2):197–213.

Merriam, J.C. 1911b. Note on a gigantic bear from the Pleistocene of Rancho La Brea. *University of California Publications, Bulletin of the Department of Geology* 6(6):163–166.

Merriam, J.C. 1912. Recent discoveries of Carnivora in the Pleistocene of Rancho La Brea. *University of California Publications, Bulletin of the Department of Geology* 7(3):39–46.

Merriam, J.C. 1912. The fauna of Rancho La Brea, Pt. 2, Canidae. *Memoirs of the University of California* 1(2):217–272.

Merriam, J.C. 1913a. The skull and dentition of a camel from the Pleistocene of Rancho La Brea. *University of California Publications, Bulletin of the Department of Geology* 7(14):305–323.

Merriam, J.C. 1913b. Preliminary report on the horses of Rancho La Brea. *University of California Publications, Bulletin of the Department of Geology* 7(21):397–418.

Merriam, J.C. 1914. Preliminary report on the discovery of human remains in an asphalt deposit at Rancho La Brea. *Science* 40:198–203.

Merriam, J.C. 1914. The brea maid. *Bulletin of the Southern California Academy of Sciences* 13(2):27–29.

Merriam, J.C. 1915. Asphalt beds of Rancho La Brea. *University of California Blue and Gold*, 15 May, 8.

Merriam, J.C. 1915. Significant features in the history of life on the Pacific coast. In *Nature and science on the Pacific coast*. San Francisco: Paul Elder and Co., 88–103.

Merriam, J.C. 1917. Felidae of Rancho La Brea. *Abstracts, Bulletin of the Geological Society of America*, (Mar.) 28:211.

Merriam, J.C. 1918a. Note on the systematic position of the wolves of the *Canis dirus* group. *University of California Publications, Bulletin of the Department of Geology* 10(27):531–533.

Merriam, J.C. 1918b. New puma-like cat from Rancho La Brea. *University of California Publications, Bulletin of the Department of Geology* 10(28):535–537.

Merriam, J.C. 1923. The cats of Rancho La Brea. *Abstracts, Journal of the Washington Academy of Sciences* 13(11):238.

Merriam, J.C. 1930. Fossils from Rancho La Brea: "A classic of science." *Science News Letter (Washington)* 18(505):378–380.

Merriam, J.C. 1930. *The living past*. New York: Charles Scribner's Sons, 144 pp.

Merriam, J.C. 1931. The cats of Rancho La Brea; a climax in evolution. *Abstracts, Science* 74:576.

Merriam, J.C., and C. Stock. 1921. Notes on peccary remains from Rancho La Brea. *University of California Publications, Bulletin of the Department of Geology* 13(2):9–17.

Merriam, J.C., and C. Stock. 1925. Relationships and structure of the short-faced bear, *Arctotherium*, from the Pleistocene of California. *Carnegie Institute of Washington Publications* 347(1):1–35.

Merriam, J.C., and C. Stock. 1932. The Felidae of Rancho La Brea. *Carnegie Institute of Washington Publications* 422:1–232.

Merriam, J.C., and C. Stock. 1933. The cats of Rancho La Brea. *Carnegie Institute of Washington, News Service Bulletin* 3(2):9–16.

Miller, A.H. 1929a. The passerine remains from Rancho La Brea in the paleontological collection of the University of California. *University of California Publications, Bulletin of the Department of Geology* 19(1):1–22.

Miller, A.H. 1929b. Additions to the Rancho La Brea avifauna. *Condor* 31:223–224.

Miller, A.H. 1937. Biotic associations and life-zones in relation to the Pleistocene birds of California. *Condor* 39:248–252.

Miller, A.H. 1940. Climatic conditions of the Pleistocene reflected by the ecologic requirements of fossil birds. *Proceedings 6th Pacific Science Congress* 2:807–810.

Miller, A.H. 1947. A new genus of icterid from Rancho La Brea. *Condor* 49(1):22–24.

Miller, G.J. 1968. On the age distribution of *Smilodon californicus* Bovard from Rancho La Brea. *Contributions in Science* 131:1–17.

Miller, G.J. 1969. A new hypothesis to explain the method of food ingestion used by *Smilodon californicus* Bovard. *Idaho State Museum, Tebiwa* 12(1):9–19.

Miller, G.J. 1969. A study of cuts, grooves, and other marks on Recent and fossil bone: I. Animal tooth marks. *Idaho State Museum, Tebiwa* 12(1):20–26.

Miller, G.J. 1969. Man and *Smilodon*: A preliminary report on their possible coexistence at Rancho La Brea. *Contributions in Science* 163:1–8.

Miller, G.J. 1970. The Rancho La Brea Project: 1969–1970. *Natural History Museum of Los Angeles County Quarterly* 9(1):26–30.

Miller, G.J. 1970. Archaeological projects: Rancho La Brea Project. *Society for California Archaeology, Newsletter* 4(2–3):11–12.

Miller, G.J. 1971. Some new and improved methods for recovering and preparing fossils as developed on the Rancho La Brea Project. *Curator* 14(4):293–307.

Miller, G.J. 1972a. Science and education at the tar pits (Part I). *Ward's Bulletin* 11(79):1–6.

Miller, G.J. 1972b. Science and education at the tar pits (Part II). *Ward's Bulletin* 11(80):1–5.

Miller, G.J. 1975. A study of cuts, grooves, and other marks on recent and fossil bones: II. Weathering cracks, fractures, splinters, and other similar natural phenomena. In *Lithic technology*, ed. E. Swanson, 211–226. The Hague: Mouton Publishers.

Miller, G.J. 1979. Some new evidence in support of the stabbing hypothesis for *Smilodon*. Abstracts, *Annual Meeting Southern California Academy of Sciences* 59:30.

Miller, G.J. 1983. Some new evidence in support of the stabbing hypothesis for *Smilodon californicus* Bovard. *Carnivore* 3(2):8–26.

Miller, G.J. 1984. On the jaw mechanism of *Smilodon californicus* Bovard and some other carnivores. *Imperial Valley College Museum Occasional Papers* 7:1–107.

Miller, G.S., Jr. 1912. The names of the large wolves of northern and western North America. *Smithsonian Miscellaneous Collections* 59(15):1–5.

Miller, L H. 1909a. *Pavo californicus*, a fossil peacock from the Quaternary asphalt beds of Rancho La Brea. *University of California Publications, Bulletin of the Department of Geology* 5(19):285–289.

Miller, L.H. 1909b. *Teratornis*, a new avian genus from Rancho La Brea. *University of California Publications, Bulletin of the Department of Geology* 5(21):305–317.

Miller, L.H. 1910a. Fossil birds from the Quaternary of southern California. *Condor* 12:12–15.

Miller, L.H. 1910b. Wading birds from the Quaternary asphalt beds of Rancho La Brea. *University of California Publications, Bulletin of the Department of Geology* 5(30):439–448.

Miller, L.H. 1910c. The condor-like vultures of Rancho La Brea. *University of California Publications, Bulletin of the Department of Geology* 6(1):1–19.

Miller, L.H. 1911a. A series of eagle tarsi from the Pleistocene of Rancho La Brea. *University of California Publications, Bulletin of the Department of Geology* 6(12):305–316.

Miller, L.H. 1911b. A synopsis of our knowledge concerning the fossil birds of the Pacific coast of North America. *Condor* 13:117–118.

Miller, L.H. 1912. Contributions to avian paleontology from the Pacific coast of North America. *University of California Publications, Bulletin of the Department of Geology* 7(5):61–115.

Miller, L.H. 1915a. The fauna of California. In *History of California*, ed. Z. S. Eldredge, 5:51–76.

Miller, L.H. 1915b. A walking eagle from Rancho La Brea. *Condor* 17:179–181.

Miller, L.H. 1916a. A review of the species *Pavo californicus*. *University of California Publications, Bulletin of the Department of Geology* 9(7): 89–96.

Miller, L.H. 1916b. The owl remains from Rancho La Brea. *University of California Publications, Bulletin of the Department of Geology* 9(8):97–104.

Miller, L.H. 1916c. Two vulturid raptors from the Pleistocene of Rancho La Brea. *University of California Publications, Bulletin of the Department of Geology* 9(9):105–109.

Miller, L.H. 1919. The walking eagle of California. *Overland Monthly* (2)70:427–429.

Miller, L.H. 1921a. A synopsis of California's fossil birds. *Condor* 23:129–130.

Miller, L.H. 1921b. Asphalt beds of Rancho La Brea. *Journal of the Washington Academy of Sciences* 11:262–263.

Miller, L.H. 1922. Fossil birds from Pleistocene of McKittrick, California. *Condor* 24:122–125.

Miller, L.H. 1923. California's ancient bird life. *University of California Chronicle*, July, 262–263.

Miller, L.H. 1924. Anomalies in the distribution of fossil gulls. *Condor* 26:173–174.

Miller, L.H. 1925. The birds of Rancho La Brea. *Carnegie Institute of Washington Publications* 349:63–106.

Miller, L.H. 1927. The falcons of the McKittrick Pleistocene. *Condor* 29:150–152.

Miller, L.H. 1928a. The antiquity of the migratory instinct in birds. *Condor* 30:119–120.

Miller, L.H. 1928b. Generic re-assignment of *Morphnus daggetti*. *Condor* 30:255–256.

Miller, L.H. 1929. The fossil birds of California. Faculty research lecture at the University of California at Los Angeles, delivered 20 May 1925. University of California at Los Angeles, 14 pp.

Miller, L.H. 1930. Dragon fly psychology. *Journal of Entomology and Zoology Pomona College*, 22: 45–46.

Miller, L.H. 1931. Pleistocene birds from the Carpinteria asphalt of California. *University of California Publications, Bulletin of the Department of Geology* 20(10):361–374.

Miller, L.H. 1932. The Pleistocene storks of California. *Condor* 34(5):212–216.

Miller, L.H. 1935. A second avifauna from the McKittrick Pleistocene. *Condor* 37:72–79.

Miller, L.H. 1940. A new Pleistocene turkey from Mexico. *Condor* 40:154–156.

Miller, L.H. 1950. *Lifelong boyhood: Recollections of a naturalist afield.* Berkeley: University of California Press, 226 pp.

Miller, L.H., and I. DeMay. 1942. The fossil birds of California: An avifauna and bibliography with annotations. *University of California Publications in Zoology* 47(4):47–142.

Miller, L.H., and H. Howard. 1938. The status of the extinct condor-like birds of the Rancho La Brea Pleistocene. *University of California at Los Angeles, Publications in Biological Sciences* 1(9): 169–176.

Miller, S.E. 1979. Pleistocene insects of Rancho La Brea and other California asphalt deposits. Abstracts, *Annual Meeting Southern California Academy of Sciences* 92:46.

Miller, S.E. 1982. Quaternary insects of the California asphalt deposits. *Proceedings 3rd North American Paleontology Convention* 2:377–379.

Miller, S.E. 1983. Late Quaternary insects of Rancho La Brea and McKittrick, California. *Quaternary Research* 20:90–104.

Miller, S.E., R.D. Gordon, and H.F. Howden. 1981. Reevaluation of Pleistocene scarab beetles from Rancho La Brea, California (Coleoptera: Scarabaeidae). *Proceedings of the Entomological Society of Washington* 83(4):625–630.

Miller, S.E., and S.B. Peck. 1979. Fossil carrion beetles of Pleistocene California asphalt deposits, with a synopsis of Holocene California Silphidae (Insecta: Coleoptera: Silphidae). *Transactions of the San Diego Society of Natural History* 19(8): 85–106.

Miller, W.E. 1968. Occurrence of a giant bison, *Bison latifrons*, and a slender-limbed camel, *Tanupolama*, at Rancho La Brea. *Contributions in Science* 147:1–9.

Miller, W.E., and J.D. Brotherson. 1979. Size variation in foot elements of *Bison* from Rancho La Brea. *Contributions in Science* 323:1–19.

Moodie, R.L. 1918. Paleontological evidences of the antiquity of disease. *Science Monthly*, Sep., 265–281.

Moodie, R.L. 1918. Studies in paleopathology. Pathological evidences of disease among ancient races of man and extinct animals. *Surgery, Gynecology and Obstetrics*, Nov., 498–510.

Moodie, R.L. 1922. On the endocranial anatomy of some Oligocene and Pleistocene mammals. *Journal of Comparative Neurology* 34:343–379.

Moodie, R.L. 1923. *Paleopathology. An introduction to the study of ancient evidences of disease.* Urbana, Illinois: University of Illinois Press. 567 pp.

Moodie, R.L. 1923. *The antiquity of disease.* Chicago: University of Chicago Press, 148 pp.

Moodie, R.L. 1926. La paléopathologie des mammifères du Pleistocene. *Biologie Médicale* 16:431–440.

Moodie, R.L. 1927. Studies in paleopathology XX. Vertebral lesions in the sabre-tooth, Pleistocene of California, resembling the so-called myositis ossificans progressiva, compared with certain ossifications in the dinosaurs. *Annals of Medical History* 9(1):91–102.

Moodie, R.L. 1928. Studies in paleodontology IV. The evidences of pyorrhea, dead teeth, and gingival infections in the mandibles of the Pleistocene giant wolf (*Aenocyon dirus*) from Rancho La Brea. *Pacific Dental Gazette* 36:414–419.

Moodie, R.L. 1929a. Excess callus in a Pleistocene bird. *American Journal of Science*, Ser. 5, 17:81–84.

Moodie, R.L. 1929b. Studies in paleodontology, XVI. The California sabre-tooth; the mandibular teeth and associated structures. *Pacific Dental Gazette* 37(6):317–321.

Moodie, R.L. 1929c. An alveolar abscess in a fossil mammal. *Pacific Dental Gazette* 37:428–433.

Moodie, R.L. 1929d. Studies in paleodontology, XX. The teeth and jaws of Nothrotherium. *Pacific Dental Gazette* 37(11):667–680.

Moodie, R.L. 1929e. Studies in paleodontology,

XXV. The California sabre-tooth, facial asymmetry following loss of sabre. *Pacific Dental Gazette* 37 (12):764–766.

Moodie, R.L. 1929f. Studies in paleodontology, XXXVII. The California sabre-tooth, Two impactions and an abscess. *Pacific Dental Gazette* 37(12):767–770.

Moodie, R.L. 1930a. Studies in paleodontology, XXII. Apical closure of root canals in adult Pleistocene carnivora. *Pacific Dental Gazette* 38(1):1–4.

Moodie, R.L. 1930b. Studies in paleopathology, XXV. Hypertrophy in the sacrum of the sabre-tooth, Pleistocene of southern California. *American Journal of Surgery* 8(6):1313–1315.

Moodie, R.L. 1930c. Studies in paleopathology, XXVI. Pleistocene luxations. *American Journal of Surgery* 9(2):348–362.

Moodie, R.L. 1930d. Studies in paleopathology, XXVII. A suggestion of rickets in the Pleistocene. *American Journal of Surgery* 10(1):162–163.

Moodie, R.L. 1930e. Studies in paleopathology, XXVIII. The phenomenon of sacralization in the Pleistocene sabre-tooth. *American Journal of Surgery* 10(3):587–589.

Moodie, R.L. 1930f. The ancient life of Yuma County. *Arizona Science Monthly* 31:401–407.

Moore, I., and S.E. Miller. 1978. Fossil rove beetles from Pleistocene California asphalt deposits (Coleoptera: Staphylinidae). *Coleoptera Bulletin* 32(1):37–39.

Nagano, C.D., S.E. Miller, and A.V. Morgan. 1982. Fossil tiger beetles (Coleoptera: Cicindelidae): Review and new Quaternary records. *Psyche* 89(3–4):339–346.

Naples, V.L. 1987. Reconstruction of cranial morphology and analysis of function in *Nothrotheriops shasternse*. *Contributions in Science* 389:1–21.

Naples, V.L. 1989. The feeding mechanism in the Pleistocene ground sloth *Glossotherium*. *Contributions in Science* 415:1–23.

Naples, V.L. 1990. Morphological changes in the facial region and a model of dental growth and wear pattern development in the Pleistocene ground sloth *Nothrotheriops shastensis. Journal of Vertebrate Paleontology* 10(3):372–389.

Nigra, J.O., and J.F. Lance. 1947. A statistical study of the metapodials of the dire wolf group from the Pleistocene of Rancho La Brea. *Bulletin of the Southern California Academy of Sciences* 46(1):26–34.

Nowak, R.M. 1979. North American Quaternary *Canis. Monographs of the Museum of Natural History, University of Kansas* 6:1–154.

Olsen, S.J. 1985. *Origins of the domestic dog: The fossil record.* Tucson: University of Arizona Press, 118 pp.

Orcutt, M.L. 1954. The discovery in 1901 of the La Brea fossil beds. *Historical Society of Southern California Quarterly* 36(4):338–341.

Ord, E.O.C. 1850. Report of Lieutenant Ord to General Riley, dated Oct. 31, 1849. In Tyson, R.T.

Report of the Secretary of War, communicating information in relation to the geology and topography of California., pp. 119–127. 31st Congress, 1st session, Senate Executive document no. 47.

Orr, P.C. 1969. *Felis trumani*, a new radiocarbon dated cat skull from Crypt Cave, Nevada. *Santa Barbara Museum of Natural History, Bulletin of the Department of Geology* 2:1–8.

Osborn, H.F. 1925. Mammals and birds of the California tar pools. Rancho La Brea and McKittrick. *Natural History*, 25:527–543.

Oswalt, S.S. 1979. Artifacts of Rancho La Brea. *Abstracts, Annual Meeting Southern California Academy of Sciences* 84:42.

Owen, P. R. 1991. The atlas of *Smilodon*. *Abstract, Annual Meeting California Academy of Sciences* no. 17.

Oxnard, C., and S. Gust. 1990. Bone biomechanics: Insights from the giant ground sloth of forty thousand years ago. *Proceedings of the Australian Society for Human Biology* 1989:12–13.

Peck, S.B., and S.E. Miller. 1980. Fossil Coleoptera from late Pleistocene asphalt deposits of southern California. *Abstracts, 16th International Congress of Entomology Kyoto, Japan*, 14.

Pierce, W.D. 1945. A case of Pleistocene myiasis from the La Brea pits. *Bulletin of the Southern California Academy of Science* 44(1):8–9.

Pierce, W.D. 1946. Exploring the minute world of the California asphalt deposits. *Bulletin of the Southern California Academy of Science* 45(3):113–118.

Pierce, W.D. 1946. Fossil arthropods of California: 11. Descriptions of the dung beetles (Scarabaeidae) of the tar pits. *Bulletin of the Southern California Academy of Sciences* 45(3):119–131.

Pierce, W.D. 1946. Fossil arthropods of California: 12. Description of a sericine beetle from the tar pits. *Bulletin of the Southern California Academy of Sciences,* 45(3):131–132.

Pierce, W.D. 1947. A progress report on the Rancho La Brea asphaltum studies. *Bulletin of the Southern California Academy of Sciences* 46(3):136–143.

Pierce, W.D. 1948. Fossil arthropods of California: 16. The carabid genus *Elaphrus* in the asphalt deposits. *Bulletin of the Southern California Academy of Sciences* 47(2):53–55.

Pierce, W.D. 1949. Fossil arthropods of California: 17. The silphid burying beetles in the asphalt deposits. *Bulletin of the Southern California Academy of Sciences* 48(2):55–70.

Pierce, W.D. 1954a. Fossil arthropods of California: 18. The Tenebrionidae-Tentyriinae of the asphalt deposits. *Bulletin of the Southern California Academy of Sciences* 53(1):35–45.

Pierce, W.D. 1954b. Fossil arthropods of California: 19. The Tenebrionidae-Scaurinae of the asphalt deposits. *Bulletin of the Southern California Academy of Sciences* 53(2):93–98.

Pierce, W.D. 1954c. Fossil arthropods of California: 20. The Tenebrionidae-Coniontinae of the asphalt deposits. *Bulletin of the Southern California Academy of Sciences* 53(3):142–156.

Pierce, W.D. 1957. Insects. *Geological Society of America Memoir* 67:943–952.

Pierce, W.D. 1961. The growing importance of paleoentomology. *Entomological Society of Washington* 63(3):211–217.

Plummer, E. 1972. Distribution crew unearths important fossil specimens. *Southern California Gas Co., Gas News* 31(8):7.

Quinn, J.P. 1991. Stratigraphic analysis of the Late Pleistocene Rancho La Brea deposits. *Abstract, Annual Meeting California Academy of Sciences* no. 3.

Raftery, P. 1979. Nonmarine ostracodes from Rancho La Brea. *Abstracts, Annual Meeting Southern California Academy of Sciences* 60:30.

Randau, J.A. 1967. Making prehistory at the La Brea tar pits. *Westways* August: 33–35.

Reynolds, R.L. 1968. Study of the human remains and associated fauna in Pit 10 of the Rancho La Brea asphalt deposits, California. *Abstracts, Annual Meeting of the Southern California Academy of Sciences.*

Reynolds, R.L. 1976. New record of *Antilocapra americana* Ord, 1818, in the late Pleistocene fauna of the Los Angeles basin. *Journal of Mammalogy* 57(1):176–178.

Reynolds, R.L. 1979. Occurrence of domestic dogs at Rancho La Brea. *Abstracts, Annual Meeting Southern California Academy of Sciences* 86:43.

Reynolds, R.L. 1985. Domestic dog associated with human remains at Rancho La Brea. *Bulletin, Southern California Academy of Sciences* 84(2):76–85.

Richards, G.D., and M.L. McCrossin. 1991. A new species of *Antilocapra* from the late Quaternary of California. *Geobios* 24 fasc. 5:623–635.

Romer, A.S. 1925. A "fossil" camel recently living in Utah. *Science* 68, no. 1749:19–20.

Romig, M.L. 1984. The resurrection of Pit 91. *Terra* 23(1):14–16.

Ruddell, M.W. 1989. Cranial variation in the dire wolf (*Canis dirus*) with a temporal framework at Rancho La Brea, Los Angeles, California. *Abstracts, Journal of Vertebrate Paleontology* (suppl. to no. 3) 9(3):37A.

Ruddell, M.W. 1991. Use of zygomatic arch width as an indicator of change through time in *Canis dirus* from Rancho La Brea, Los Angeles, California. *Abstract, Annual Meeting California Academy of Sciences* no. 14.

Salls, R.A. 1980. The La Brea cogged stone. *Masterkey, Southwest Museum* 54(2):53–59.

Salls, R.A. 1986. The La Brea atlatl foreshafts: Inferences for the Millingstone Horizon. *Pacific Coast Archaeological Society Quarterly* 22(2):21–30.

Salls, R.A. 1991. Faunal remains and social status: An analysis of human repast in a tar pit. *Abstract, Annual Meeting California Academy of Sciences* no. 22.

Savage, D.E. 1951. Late Cenozoic vertebrates of

the San Francisco Bay region. *University of California Publications in Geological Sciences* 28(10): 215–314.

Savage, D.E., T. Downs, and O.J. Poe. 1954. Cenozoic land life of southern California. In *Geology of southern California*, 53–57. *Bulletin of the California Division of Mines* no. 170.

Savage, J.M. 1963. Studies on the lizard family Xantusiidae IV. The genera. *Contributions in Science* 71:1–38.

Schultz, G.E., and C.H. Lansdown. 1972. A skull of *Bison latifrons* from Lipscomb County, Texas. *Texas Journal of Science* 23(3):393–401.

Schultz, J.R. 1938. A late Quaternary mammal fauna from the tar seeps of McKittrick, California. *Carnegie Institute of Washington Publications* 487:118–161.

Scott, E. 1985. They live again: Sixty years of sculpture in Hancock Park. *Terra* 24(1):23–29.

Scott, E. 1986. Out of the past: Hancock Park statuary brings prehistoric beasts to life. *Herald Examiner, California Living*, 27 April, 18–21.

Scott, E. 1988. Mount up! The skeletons of Rancho La Brea on display. *Terra* 26(6):11–13.

Scott, E. 1989. Skeletal remains of *Equus* from the Page Museum salvage, Rancho La Brea: A preliminary report. *Current Research in the Pleistocene* 6:78–81.

Scott, E. 1990. New record of *Equus conversidens* Owen, 1869 (Mammalia; Perissodactyla; Equidae) from Rancho La Brea. *Journal of Vertebrate Paleontology* supplement to 10(3):41A.

Scott, E. 1991a. Ontogenetic age and sex distributions of large *Equus* (Mammalia; Perissodactyla; Equidae) from Rancho La Brea, Los Angeles, California. *Abstract, Annual Meeting California Academy of Sciences* no. 11.

Scott, E. 1991b. Was La Brea Woman a mother? *Abstract, Annual Meeting California Academy of Sciences* no. 20.

Scott, E., and S. Gust. 1990. Sexual dimorphism in fossil populations: The Rancho La Brea horse through time. *Abstracts, PaleoBios* (suppl. to no. 49) 13(49):8.

Scott, E., R.L. Reynolds, D.E. Bleitz, and R.A. Salls. 1991. Prehistoric artifacts from Rancho La Brea. *Abstract, Annual Meeting California Academy of Sciences* no. 21.

Scott, F.M. 1972. Stallate epidermal hairs, some 10,000 years old. *Madroño* 21(7):458.

Seegmiller, R.F., and R.D. Ohmart. 1981. Ecological relationships of feral burros and desert bighorn sheep. *Wildlife Monographs* 78:1–58.

Shackleford, J.M., and R.W.G. Wyckoff. 1964. Collagen in fossil teeth and bones. *Journal of Ultrastructure Research* 11:173–180.

Shaw, C.A. 1979. Techniques used in excavation of LACM 6909 (Pit 91), Rancho La Brea. *Abstracts, Annual Meeting Southern California Academy of Sciences* 58:29.

Shaw, C.A. 1982. Techniques used in excavation, preparation, and curation of fossils from Rancho La Brea. *Curator* 25(1):63–77.

Shaw, C.A. 1988. Body by Fischer and Bessom. *Terra* 26(6):13–16.

Shaw, C.A. 1989. The collection of pathologic bones at the George C. Page Museum, Rancho La Brea, California: A retrospective view. *Abstracts, Journal of Vertebrate Paleontology*, (suppl. to no. 3) 9(3):38A.

Shaw, C.A., F.P. Heald, and M.L. Romig. 1991. Paleopathological evidence of social behavior in *Smilodon fatalis* from Rancho La Brea. *Abstract, Annual Meeting California Academy of Sciences* no. 19.

Shaw, C.A., and J.P. Quinn. 1986. Rancho La Brea: A look at coastal southern California's past. *California Geology* 39(6):123–133.

Shaw, C.A., and A.E. Tejada-Flores. 1985. Biomechanical implications of the variation in *Smilodon* ectocuneiforms from Rancho La Brea. *Contributions in Science* 359:1–8.

Shermis, S. 1979. The La Brea woman. *Abstracts, Annual Meeting Southern California Academy of Sciences* 85:43.

Shermis, S. 1983. Healed massive pelvic fracture in a *Smilodon* from Rancho La Brea. *Paleobios* 1(3):121–126.

Shermis, S. 1984. Healed massive pelvic trauma in a *Smilodon* from Rancho La Brea. *Abstracts, Annual Meeting Southern California Academy of Sciences* 37:18.

Shermis, S. 1985. Alveolar osteitis and other oral diseases in *Smilodon californicus*. *Ossa* 12:187–196.

Shermis, S. 1985. Canine fracture avulsion in *Smilodon californicus*. *Bulletin of the Southern California Academy of Sciences* 84(2):86–95.

Sibley, C. 1939a. Fossil fringillids from Rancho La Brea. *Condor* 41:126–127.

Sibley, C. 1939b. Chipping sparrows in the Rancho La Brea. *Condor* 41:258–259.

Sibley, G. 1950. A very early resident. *Natural History Museum of Los Angeles County Quarterly* 8(2):13–14.

Sibley, G. 1967. *La Brea story*. Los Angeles: Natural History Museum of Los Angeles County, Education Division, 46 pp.

Sibley, G. 1977. A new home for Rancho La Brea fossils. *Ward's Bulletin for Biology, Earth Sciences and Chemistry* 17(105):

Simpson, L.B., ed. and trans. 1961. *Journal of José Longinos Martínez: 1791-1792*. San Francisco: John Howell-Books, 114 pp.

Simpson, G.G. 1941. Large Pleistocene felines of North America. *American Museum Novitates* 1136:1–26.

Simpson, G.G. 1951. Chester Stock. *National Academy Biographical Memoirs* 27:335–362.

Sinclair, W.J. 1910. Dermal bones of *Paramylodon* from the asphaltum deposits of Rancho La Brea, near Los Angeles, California. *Proceedings of the American Philosophical Society* 49:191–195.

Slaughter, R.H. 1961. A new coyote in the late

Pleistocene of Texas. *Journal of Mammalogy* 42: 503–509.

Slaughter, R.H. 1966. *Platygonus compressus* and associated fauna from the Laubach Cave of Texas. *American Midland Naturalist* 75:475–494.

Snure, H. 1924. A roentgen-ray study of the La Brea (Calif.) fossils. *American Journal of Roentgenology and Radium Therapy* 11(4):351–354.

Snyder, C.T., G. Hardman, and F.F. Zdenek. 1964. Pleistocene lakes in the Great Basin. *United States Geological Survey Miscellaneous Geologic Investigations, Map I-416.*

Soper, E.K. 1943. Salt Lake oil field. In *Geology of California—the occurrence of oil and gas,* ed. O.P. Jenkins, 284–286. *California Division of Mines Bulletin* 118(8).

Sperling, J.A. 1991. Diatoms of Rancho La Brea. *Abstract, Annual Meeting California Academy of Sciences* no. 4.

Steadman, D.W. 1980. A review of the osteology and paleontology of turkeys (Aves: Meleagridinae). In *Papers in avian paleontology honoring Hildegarde Howard,* ed. K.E. Campbell, Jr., 131–207. *Contributions in Science* 330.

Steadman, D.W., and P.S. Martin. 1984. Extinction of birds in the late Pleistocene of North America. In *Quaternary extinctions: A prehistoric revolution,* ed. P.S. Martin and R.G. Klein, 466–477. Tucson: University of Arizona Press.

Steinmetz, J.J. 1991. Ostracodes of the Rancho La Brea deposits. *Abstract, Annual Meeting California Academy of Sciences* no.6.

Stirton, R.A. 1938. Notes on some late Tertiary and Pleistocene antilocaprids. *Journal of Mammalogy* 19(3):336–370.

Stock, C. 1913. *Nothrotherium* and *Megalonyx* from the Pleistocene of Southern California. *University of California Publications, Bulletin of the Department of Geology* 7(17):341–358.

Stock, C. 1914a. The systematic position of the mylodont sloths from Rancho La Brea. *Science* 39:761–763.

Stock, C. 1914b. Skull and dentition of the mylodont sloths of Rancho La Brea. *University of California Publications, Bulletin of the Department of Geology* 8(18):319–334.

Stock, C. 1917a. Recent studies on the skull and dentition of *Nothrotherium* from Rancho La Brea. *University of California Publications, Bulletin of the Department of Geology* 10(10):137–164.

Stock, C. 1917b. Further observation on the skull structure of mylodont sloths from Rancho La Brea. *University of California Publications, Bulletin of the Department of Geology* 10(11):165–178.

Stock, C. 1917c. Structure of the pes in *Mylodon harlani. University of California Publications, Bulletin of the Department of Geology* 10(16): 267–286.

Stock, C. 1920a. A mounted skeleton of *Mylodon harlani. University of California Publications, Bulletin of the Department of Geology* 12(6):425–430.

Stock, C. 1920b. Origin of the supposed human footprints of Carson City, Nevada. *Science* 51: 514.

Stock, C. 1925. Cenozoic gravigrade edentates of western North America with special reference to the Pleistocene Megalonychinae and Mylodontidae of Rancho La Brea. *Carnegie Institute of Washington Publications* 331:1–206.

Stock, C. 1928. A peccary from the McKittrick Pleistocene, California. *University of California Publications, Bulletin of the Department of Geology* 393:25–27.

Stock, C. 1929a. Significance of abraded and weathered mammalian remains from Rancho La Brea. *Bulletin of the Southern California Academy of Sciences* 28(1):1–5.

Stock, C. 1929b. A census of the Pleistocene mammals of Rancho La Brea, based on the collections of the Los Angeles Museum. *Journal of Mammalogy* 10(4):281–289.

Stock, C. 1930. *Rancho La Brea: A record of Pleistocene life in California.* Science Series, no. 1. Los Angeles: Los Angeles Museum, 84 pp.

Stock, C. 1932. Asphalt deposits and Quaternary life of Rancho La Brea. *Southern California XVI International Geological Congress, Guide-book* 15:21–23.

Stock, C. 1936. *Ursus,* or the past of the California bears. *Westways,* 28 Nov., no. 11, 30.

Stock, C. 1937. California buffalo of long ago. *Westways,* 29 Feb., no. 2, 29.

Stock, C. 1938. A coyote-like wolf-jaw from the Rancho La Brea Pleistocene. *Bulletin, Southern California Academy of Sciences* 37(2):49–51.

Stock, C. 1941. Prehistoric archeology. In *Geology, 1888–1938, 50th anniversary volume Geological Society of America,* 139–158.

Stock, C. 1942. *Rancho La Brea: A record of Pleistocene life in California.* Rev. ed. Science Series no. 4, Paleontology Publication no. 4. Los Angeles: Los Angeles County Museum, 73 pp.

Stock, C. 1944. California bears, present and past. *Engineering and Science Monthly* 7(7):12–14.

Stock, C. 1946. *Rancho La Brea: A record of Pleistocene life in California.* 3d ed. Science Series no. 11, Paleontology Publication no. 7. Los Angeles: Los Angeles County Museum, 74 pp.

Stock, C. 1949. *Rancho La Brea: A record of Pleistocene life in California.* 4th ed. Science Series no. 14, Paleontology Publication no. 8. Los Angeles: Los Angeles County Museum, 81 pp.

Stock, C. 1956. *Rancho La Brea: A record of Pleistocene life in California.* 6th ed. Science Series no. 20, Paleontology Publication no. 11. Los Angeles: Natural History Museum of Los Angeles County, 81 pp.

Stock, C., and J.F. Lance. 1948. The relative lengths of limb elements in *Canis dirus. Bulletin of the Southern California Academy of Sciences* 47(3): 79–83

Stock, C., J.F. Lance, and J.O Nigra. 1946. A newly mounted skeleton of the extinct dire wolf from the Pleistocene of Rancho La Brea. *Bulletin of the Southern California Academy of Sciences* 45(2):108–110.

Stoner, R.C. 1913. Recent observations on the mode of accumulation of the Pleistocene bone deposits of Rancho La Brea. *University of Cali-*

fornia Publications, Bulletin of the Department of Geology 7(20):387–396.

Sushkin, P.P. 1928. On the affinities of *Pavo californicus*. *Ibis,* Jan., 135–138.

Swarth, H.S. 1915. Guide to the exhibit of fossil animals from Rancho La Brea. *Los Angeles Museum of History, Science and Art, Miscellaneous Publications* 1:1–34.

Swift, C.C. 1979. Freshwater fish of the Rancho La Brea deposit. *Abstracts, Annual Meeting Southern California Academy of Sciences* 88:44.

Swift, C.C. 1989. Freshwater fishes from the Rancho La Brea deposit, Southern California. *Bulletin of the Southern California Academy of Sciences* 88(3):93–102.

Taylor, W.P. 1911. A new antelope from the Pleistocene of Rancho La Brea. *University of California Publications, Bulletin of the Department of Geology* 6(10):191–197.

Tejada-Flores, A.E. 1979. *Smilodon* hyoids from Rancho La Brea. *Abstracts, Annual Meeting Southern California Academy of Sciences* 91:46.

Tejada-Flores, A.E., and C.A. Shaw. 1984. Tooth replacement and skull growth in *Smilodon* from Rancho La Brea. *Journal of Vertebrate Paleontology* 4(1):114–121.

Templeton, B.C. 1955. Fossil plants in the La Brea deposits. *Natural History Museum of Los Angeles County Quarterly* 12(1):8–11.

Templeton, B.C. 1964. The fruits and seeds of the Rancho La Brea Pleistocene deposits. Ph.D. diss., Oregon State University, Corvallis, Oregon, 224 pp.

Templeton, B.C. 1964. The fruits and seeds of the Rancho La Brea Pleistocene deposits. *Dissertation Abstracts* 25:3228–3229.

Valentine, V.W., and J.H. Lipps. 1970. Marine fossils at Rancho La Brea. *Science* 169:277–278.

Van Valkenburgh, B.F. Hertel, and W. Anyonge. 1991. Feeding behavior of *Smilodon*: Evidence from dental microwear and tooth fracture frequencies. *Abstract, Annual Meeting California Academy of Sciences* no. 18.

Van Vuren, D. 1984. Summer diets of bison and cattle in southern Utah. *Journal of Range Management* 37:260–261.

vonBloeker, J.C., Jr. 1944. New locality records for some west American shrews. *Journal of Mammalogy* 25:311–312.

Waldo, E. 1958. Wildlife of yesteryear. *Louisiana Conservationist Wildlife Bulletin* 68:1–7.

Wallace, W.J. 1955. A suggested chronology for southern California coastal archaeology. *Southwest Journal of Anthropology* 11:214–230.

Wallace, W.J. 1971. A suggested chronology for southern California coastal archaeology. In *The California Indians: A source book,* ed. R.F. Heizer and M.A. Whipple, 186–201. Berkeley: University of California Press.

Warter, J.K. 1976. Late Pleistocene plant communities—evidence from the Rancho La Brea tar pits. *Symposium Proceedings on Plant Communities of Southern California, Special Publications of California Native Plant Society* 2:32–39.

Warter, J.K. 1979. The environment of Pit 91, Rancho La Brea, as interpreted by plant remains. *Abstracts, Annual Meeting Southern California Academy of Sciences* 87:44.

Warter, J.K. 1980. Late Pleistocene environment of the Los Angeles basin based on plant remains from the Rancho La Brea tar pits. *Abstracts, Bulletin of the Ecological Society of America* 61(2):107.

Webb, S.D. 1965. The osteology of *Camelops*. *Natural History Museum of Los Angeles County, Science Bulletin* 1:1–54.

Webb, S.D. 1973. Pliocene pronghorns of Florida. *Journal of Mammalogy* 54(1):203–221.

Webb, S.D. 1974. Pleistocene llamas of Florida with a brief review of the Lamini. In *Pleistocene mammals of Florida,* ed. S.D. Webb, 170–213. Gainesville: University Presses of Florida,.

Webster, R. 1979. Birding the tar pits—12,000 B.C. *Los Angeles Audubon Society* 45(10):1–5.

Wetmore, A. 1924. Fossil birds from southeastern Arizona. *Proceedings of the United States National Museum* 64, (5):1–18.

Wetmore, A. 1927. Present status of the checklist of fossil birds for North America. *Auk* 44:179–183.

Wetmore, A. 1928a. Prehistoric ornithology in North America. *Journal of the Washington Academy of Sciences* 18:145–158.

Wetmore, A. 1928b. Birds of the past in North America. *Smithsonian Institution Annual Report* 1928:377–390.

Whistler, D.P. 1989. Late Pleistocene chipmunk, *Tamias* (Mammalia: Sciuridae), from Rancho La Brea, Los Angeles, California. *Bulletin of the Southern California Academy of Sciences* 88(3):117–122.

Willoughby, D.P. 1948. A statistical study of the metapodials of *Equus occidentalis* Leidy. *Bulletin of the Southern California Academy of Sciences* 47:84–94.

Willoughby, D.P. 1974. *The empire of Equus.* South Brunswick and New York: A. S. Barnes and Co., 475 pp.

Wilson, R.W. 1933. Pleistocene mammalian fauna from the Carpinteria asphalt. *Carnegie Institute of Washington Publications* 440(6):59–76.

Wilson, M.C. 1991. The La Brea bison: Dentitions, population dynamics, and taxonomy. *Abstract, Annual Meeting California Academy of Sciences* no. 12.

Winans, M.C. 1985. Revision of North American fossil species of the genus *Equus* (Mammalia: Perissodactyla: Equidae). Ph.D. diss., University of Texas at Austin, 265 pp.

Winans, M.C. 1989. A quantitative study of North American fossil species of the genus *Equus.* In *The evolution of the Perissodactyls,* ed. D.R. Prothero and R.M. Schoch, 263–297. New York: Oxford University Press.

Winans, M.C., and R.C. Winans. 1982. Measuring systems, techniques, and equipment for taphonomic studies. *University of California Los Angeles, Archaeological Research Tools, Institute of Archaeology* 2:45–58.

Woodard, G.D., and L.F. Marcus. 1971. Late Pleistocene stratigraphy, Rancho La Brea fossil deposits, Los Angeles, California. *Abstracts, Geological Society of America* 3(2):218.

Woodard, G.D., and L.F. Marcus. 1973. Rancho La Brea fossil deposits: A re-evaluation from stratigraphic and geological evidence. *Journal of Paleontology* 47(1):54–69.

Woodard, G.D., and L.F. Marcus. 1976. Reliability of late Pleistocene correlation using C-14 dating: Baldwin Hills–Rancho La Brea, Los Angeles, California. *Journal of Paleontology* 50(1):128–132.

Woodring, W.P., M.N. Bramlette, and W.S.W. Kew. 1946. Geology and paleontology of Palos Verdes Hills California. *U.S. Geological Survey, Professional Paper* no. 207:1–145.

Woodward, A. 1937. Atlatl dart foreshafts from the La Brea pits. *Bulletin of the Southern California Academy of Sciences* 36:41–59.

Wyckoff, R.W.G. 1964. Proteins from Rancho La Brea fossils. *Natural History Museum of Los Angeles County Quarterly* 2(4):11.

Wyckoff, R.W.G. 1972. *The biochemistry of animal fossils*. Bristol: Scientechnica Publications Ltd., and Baltimore: Williams and Wilkins Co., 152 pp.

Wyckoff, R.W.G., and A.R. Doberenz. 1965. The electron microscopy of Rancho La Brea bone. *Proceedings of the National Academy of Sciences* 53(2):230–233.

Wyckoff, R.W.G., and A.R. Doberenz. 1968. The strontium content of fossil teeth and bones. *Geochimica Cosmochimica Acta* 32:109–115.

Wyckoff, R.W.G., W.F. McCaughey, and A.R. Doberenz. 1964. The amino acid composition of proteins from Pleistocene bones. *Biochimica Biophysica Acta* 93:374–377.

Wyckoff, R.W.G., E. Wagner, P. Matter III, and A.R. Doberenz. 1963. Collagen in fossil bone. *Proceedings of the National Academy of Sciences* 50(2):215–218.

Wyman, L.E. 1918. *Notes on the Pleistocene fossils obtained from Rancho La Brea asphalt pits.* Los Angeles Museum, Miscellaneous Publications, no. 2, 35pp.

Wyman, L.E. [1918] 1922. *Notes on the Pleistocene fossils obtained from Rancho La Brea asphalt pits.* Reprint. Los Angeles Museum, Miscellaneous Publications, no. 2, 35pp.

Wyman, L.E. [1918] 1926. *Notes on the Pleistocene fossils obtained from Rancho La Brea asphalt pits.* Reprint. Los Angeles Museum, Miscellaneous Publications, no. 2, 35pp.

Wyman, L.E. 1926. Museum sketches: An unwieldy exhibit explained. *Los Angeles Museum, Museum Graphic* 1(1):32–34.

Wyman, L.E. 1927. La Brea in retrospect. *Los Angeles Museum, Museum Graphic* 1(3):82–87.

Young, F.B., and A.L. Cooper. 1926. Evidence of diseases as shown in fossil and prehistoric remains; paleopathology. *Transactions of the Section of Pathology and Physiology, American Medical Association* 1:1–11.

Young, F.B., and A.L. Cooper. 1927. A study in paleopathology. *Radiology* 8:230–240.

Zechmeister, L., and W. Lijinsky. 1953. Some neutral constituents of a natural tar originating from the La Brea pits. *Archives of Biochemistry and Biophysics* 47(2):391–395.

Zhao, Y., D.R. Boone, R.A. Mah, J.E. Boone, and L. Xun. 1989. Isolation and characterization of *Methanocorpusculum labreanum* sp. nov. from the La Brea tar pits. *International Journal of Systematic Bacteriology* 39(1):10–13.

INDEX

Page numbers in *italics* refer to figures.